THE TI-83® COMPANION

to accompany

ELEMENTARY STATISTICS

SEVENTH EDITION

THE TI-83® COMPANION

LARRY A. MORGAN
Montgomery County Community College

to accompany

ELEMENTARY STATISTICS
SEVENTH EDITION
MARIO F. TRIOLA

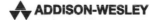 **ADDISON-WESLEY**

An imprint of Addison Wesley Longman, Inc.

Reading, Massachusetts • Menlo Park, California • New York • Harlow, England
Don Mills, Ontario • Sydney • Mexico City • Madrid • Amsterdam

To Mother, Abui, Shaine, and Cheyenne

Reprinted with corrections, July 1998

Cover illustration by Dave Cutler.

Reproduced by Addison-Wesley Publishing Company Inc. from camera-ready copy supplied by the author.

ISBN   0-201-30727-8

2 3 4 5 6 7 8 9 10 MA 009998

Preface

The TI-83 is a little computer with many capabilities. This companion will introduce you to some of its statistical capabilities as they relate to *Elementary Statistics* (7th ed.) by Mario F. Triola. This companion is designed to be used side by side with the text. For example, Chapter 1 of *Elementary Statistics* has a calculator icon by the definition of a *simple random sample* [pg. 19] and one by Exercise 18 [pg. 25] requesting such a sample. These icons indicate that the procedure is carried out in detail in Chapter 1 of this companion. The definition, exercise number, and text page number (in square brackets) provide cross references from this companion back to the main book.

Although this companion is designed primarily for the TI-83, it also includes suggestions for the TI-82. When the TI-83 has a built-in function that is not on the TI-82, an alternate procedure is given. This often requires doing a calculation on the TI-82 and then using the tables in the text. The advanced topics of multiple regression in Chapter 9 and ANOVA in Chapter 11 rely on computer output to cover the material in the text. The TI-83, but not the TI-82, have programs that can duplicate the text's computer output. You can obtain these programs from your instructor or from a disk that is available from the publisher. Also included on the disk are all the data sets referenced in Appendix B of the text. The appendix of this companion explains how you can download these programs and data to your TI-83 from a computer or another TI-83.

This companion begins with a brief section, "Introduction to the TI-83," that introduces the TI-83 keyboard and notation. After that, it follows the text of *Elementary Statistics* (7th ed.) chapter by chapter, offering helpful techniques on the TI-83. Most of the examples and data are from the text; these examples are paraphrased but should be complete enough for continuity. Consult the text for important details such as the appropriateness of the procedure used.

Even though the TI-83 can often perform operations in more than one way, I do not show all the possibilities but concentrate instead on the method that will be most valuable in the long run.

The "Introduction to the TI-83" section, Chapter 1, and Chapter 2 are very important for your understanding of all the chapters that follow. Read them carefully, and refer to them as often as necessary to remember the meanings of such phrases as "Home screen" and "the last entry feature," and the ways to perform such procedures as saving and deleting data. You will need to understand these terms and procedures before you use the material in later chapters.

The back cover of this companion has a "TI-83 Quick Reference" that summarizes some important keys, menus, and functions used in this companion.

Although this is a small book, I had a large amount of help with it. I thank my students and colleagues who have supported the use of graphic calculators in statistics classes, with special thanks to Roseanne Hofmann. Thanks to Charlotte Andreini of Texas Instruments. Thanks, of course, to Mario F. Triola, who made this companion possible. Thanks to Julia Berrisford, Susan London-Payne, Sally Stickney, Carolyn Lee, and the staff of Addison Wesley Longman, Inc., who helped turn my rough draft into a useful companion. Thanks also to Abui and Bernie for their continued support.

Contents

THE TI-83® COMPANION

to accompany

ELEMENTARY STATISTICS

SEVENTH EDITION

Introduction to the

TI-83

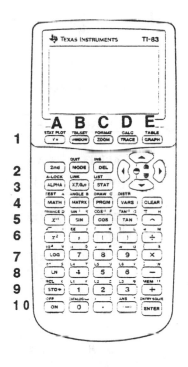

The TI-83 graphics calculator is in reality a hand-held computer and an enhanced TI-82. Many of the enhancements make it even better for calculating statistics. The TI-82 will be mentioned specifically in this companion only when it is necessary to distinquish it from the TI-83. This introduction will provide an overview of the TI-83 keyboard and give the notation used for the various keys and menus. Read this section carefully so you will know which keys are being referred to throughout this companion.

In this section you will also learn how to set the correct MODE on the TI-83 to ensure that you will obtain the same results as this companion does. And so you do not end up squinting at the screen trying to read it as it gets ever dimmer, you will find out how to adjust the screen contrast to keep it sharp and how to check the battery strength. Also included in this section are some important menus that will be used throughout. You can skim this material for now, but you will be referring to it later as needed.

KEYBOARD AND NOTATION

The TI-83 keyboard has 5 columns (designated in the picture above as A, B, C, D, E) and 10 rows of keys. The cursor control, or arrow, keys (▲ ▼ ◀ ▶) toward the upper right of the keyboard disturb the pattern but in a logical way. (See the keyboard schematic for the TI-83 above.)

Start in the upper left corner with the **Y=** key in the A column and the first row, or the A1 location. (Touch the keys mentioned as you follow along.) The **ON** key is at A10, **ENTER** is at E10, and the **GRAPH** key is at E1. These four keys are the corners of the TI-83 keyboard.

The alphabetized Key Lookup Table, on the next page, will be helpful in finding keys mentioned in this companion. After each key name is its location on the keyboard.

KEY LOOKUP TABLE	
Key	A t
2nd	A2
ALPHA	A3
↑ ANS	D10
CLEAR	E4
DEL	C2
* ↑ DISTR	D4
ENTER	E10
↑ ENTRY	E10
GRAPH	E1
↑ INS	C2
↑ LIST	C3
MATH	A4
MATRX	B4
↑ MEM	E9
MODE	B2
ON	A10
PRGM	C4
↑ QUIT	B2
STAT	C3
↑ STAT PLOT	A1
STO►	A9
TRACE	D1
VARS	D4
WINDOW	B1
Y =	A1
ZOOM	C1
÷ for /	E6
x for *	E7

* Not on the TI-82

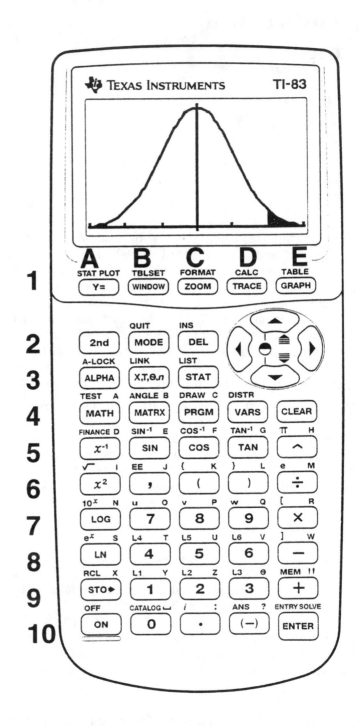

Many keys on the TI-83 have multiple functions. The primary function is marked on the key itself, and other functions are marked in colors above the key. Let's see how these functions marked in colors work.

Yellow 2nd Key at A2

The **STAT** key is at C3; above it, in yellow (blue on the TI-82), is ↑**LIST**, which is engaged by pressing and releasing the yellow **2nd** key and then the **STAT** key. (You know the **2nd** key has been engaged when the cursor key is a blinking up arrow.) Other such key combinations will be for ↑**QUIT**, ↑**INS**, ↑**ANS**, and ↑**ENTRY**.

In this manual a small up arrow before a word designates that the word is in yellow and thus you must hit the yellow **2nd** key first. You must also press the **2nd** key first with the yellow list designations **L1**, **L2**, . . . , **L6** above the numerals **1**, **2**, . . . , **6** and with the yellow { } set, or list braces, above the regular parenthesis keys ( ) at C6 and D6. No up arrow is used for these, however, because their locations are easy to remember.

Green ALPHA Key at A3

You can use the letters of the alphabet to assign data storage locations and to name a list. To engage them, first press and release the **ALPHA** key. For example, the letter **B** is engaged by hitting **ALPHA** and then pressing the **MATRX** key at B4, which has the letter **B** above it to the right in green (gray on the TI-82). You know the **ALPHA** key is engaged when the cursor is a blinking **A**.

Some General Keyboard Patterns and Important Keys

1. The first row is for plotting and graphing.
2. The second row has the important **2nd** ↑**QUIT** combination, the **DEL** ↑**INS** key, and the cursor control keys, which are all used for editing. Just below the cursor control keys is the **CLEAR** key at E4.
3. The A column has the **MATH** key and various math functions, for example, x^2, $\sqrt{\ }$.
4. The E column has the math operations $+$, $-$, $\times$, $\div$, $\wedge$.

    Note: The $\div$ shows on the TI-83 screen as **/** and the $\times$ as $*$.

5. The **STAT** ↑**LIST** key at C3 tops a pyramid of important keys based in row 4: **MATRX**, **PRGM**, and **VARS** ↑**DISTR**. (↑**DISTR** is not on the TI-82.)
6. The sixth row, above the numeral rows, has ({ , }), which are used for grouping and spacing.
7. The **STO▶** key at A9 is used for storing. It shows as a → on the display screen.
8. The tenth row has the negative symbol (-) ↑**ANS** key, which differs from the subtraction key at E8, and the biggest key, the **ENTER** ↑**ENTRY** key.

    Note: The (-) shows as ⁻ on the screen—smaller and higher than the subtraction sign.

SETTING THE CORRECT MODE AT B2

If your answers do not show as many decimal places as the examples in this companion have, or if you have difficulty matching other output, check your **MODE** settings.

From the Home screen, the screen that appears when you turn on your TI-83, press **MODE** at B2. The menu on the right appears, with the first word in each row darkened. If your screen looks different, use the cursor control key to go to each divergent row and then, with the first element blinking, press **ENTER**. Repeat this procedure until the screen looks like the one at the right.

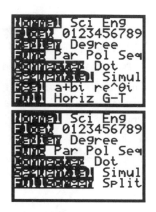

Engage ↑**QUIT** at B2 to return to the Home screen.

Note: The TI-83 mode screen at the top differs from the lower TI-82 screen only in the last two lines. The menu screen of the TI-83 and the TI-82 also have some differences but these differences will not be mentioned unless they affect the procedure being discussed.

Note: All other settings are assumed to be factory or default settings and may be restored as explained on pages 18-5 and 18-6 of the Guidebook that comes with the TI-83.

SCREEN CONTRAST ADJUSTMENT AND BATTERY CHECK

To adjust the screen contrast, follow these steps:

From the Home screen, press and release the **2nd** key and hold down the ▲ cursor control key to increase the contrast. Notice that the number in the upper right corner of the screen increases from 0 (lightest) to 9 (darkest). Press and release the **2nd** key and hold down the ▼ key to lighten the contrast. (A visual reminder of this appears on the TI-83 with the yellow and black circle between the cursor keys on the keypad.) If you adjust the contrast setting to 0, the display may become completely blank. Press and release the **2nd** key and then hold down the ▲ key until the display reappears.

When the batteries are low, the display begins to dim (especially during calculations), and you must adjust to a higher contrast setting. If you have to set the contrast setting to 9, you will need to replace the four AAA batteries soon. (If your batteries are low, the TI-83 displays a low-battery message when you turn on the calculator.) The display contrast may appear very dark after you change batteries. Press and release the **2nd** key and then hold down the ▼ key to lighten the display.

IMPORTANT MENUS

Some of the important menus you will be encountering repeatedly in your statistical work are listed here and in a briefer form on the back cover. The key that calls the menus is given above it.

Note: These are TI-83 menus; TI-82 menus will look a bit different.

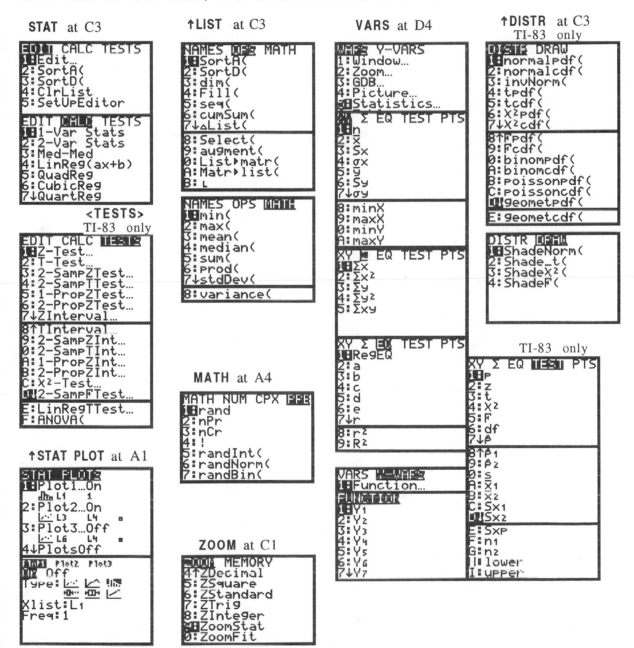

1

Introduction
to Statistics

```
L1        BLUE      TAR       2
  10      .838      16
  20      .875      16
  30      .87        9
  40      .956       8
  50      .968      16
  60      -----     13
  70                15
BLUE(5) = .968
```

In this chapter we begin our study of statistics with Triola's *Elementary Statistics* (7th ed.) and with the TI-83 by taking an exercise from the text and showing you how to select a simple random sample. The example also introduces the notation that will be used throughout this companion when menu items are selected. You will then learn how to do basic calculations on the Home screen. The Home screen keeps past entries in sight so that you can edit or modify them and then store them for future use. You will then see how lists of data can be saved and edited in the STAT editor or spreadsheet.

In this chapter, as in the chapters that follow, we do not summarize all the important text information. Our purpose in this companion is to help you use the TI-83 in conjunction with the text. Be sure to study the text for a full discussion of definitions, concepts, and proper statistical procedures.

To understand the material in this chapter, you will need to have read the preceding "Introduction to the TI-83" section. And to help you become comfortable using your graphing calculator, we suggest that you follow along with each example by keying the instructions given into your own TI-83.

RANDOM SAMPLES AND MENU NOTATION

DEFINITION [pg. 19]: . . .(A *simple random sample* of n subjects is selected in such a way that every possible sample of size n has the same chance of being chosen.) . . . . Random samples have been selected with methods such as using computers to generate random numbers or using tables of random numbers.

Note: The numbers in brackets refers to the pages in *Elementary Statistics*.

EXERCISE 18 [pg. 25]: Describe in detail a method that could be used to obtain a simple random sample of the heights of five students from a statistics class.

1. Give each student a different consecutive counting number starting from 1 as an ID number. For example, you would use 1 to 28 for a class of 28 students.

2. Press the **ON** key on the TI-83. A cursor should be blinking on the Home screen. If not, press ↑**QUIT**, at B2, to return to the Home screen.

3. Press the **CLEAR** key if the cursor is not in the upper left corner.

Note: **CLEAR**, at E4, clears a line to the left *or* the screen above it.

4. We will generate five random numbers from 1 to 28. The numbers are random in the sense that each has an equal chance of being generated; but they are generated mathematically so that you can get the same results as shown below by setting the same seed as follows.

(a) On the Home screen, type **123**; then press **STO►**, then **MATH**, then ◄ , and then the left cursor control key *once* (or the right cursor control key *three* times), to highlight **PRB** for the menu shown in screen (1).

(1)

(b) Then press **1**, and **rand** is pasted to the Home screen, as shown in the first line in screen (2).

(c) Press **ENTER**, and the **123** on the second line of screen (2) indicates that the seed is set.
Note: In the future this type of sequence of steps will be given as 123 **STO► MATH** <PRB> 1:rand or 123 **STO►**rand.

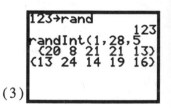

(2)

TI-82 users read step 5 and continue with step 6.

5. Press **MATH** <PRB> 5:randInt(, and then type **1,28,5** (be sure to type commas); then press **ENTER** for the fourth line of screen (3) with 20, 8, 21, and 13 being the first four random IDs. Since 21 was repeated, press **ENTER** again; 24 becomes the fifth random ID.

Note: If we used randInt(**1,28,10** to have backups for possible repeats, the set of numbers would not fit on one line and we could use ► key to reveal the unseen numbers.

(3)

6. Measure the heights of the students with ID numbers 20, 8, 21, 13, and 24.

7. To show how randInt works internally (and to show how the above could be done on the TI-82), press ↑**ENTRY** at E10; the <u>last entry</u> randInt(1,28,5 returns to the screen. Press ↑**ENTRY** again, and the entry before that, or 123 → rand, returns to the screen. Press **ENTER**, and the seed is reset as above and as in the first two lines in screen (4).

8. Press **MATH** <PRB> 1:rand**×28+1** for the third line in screen (4). Press **ENTER** four times for the first three IDs of 20, 8, and 21 (ignoring the decimals).

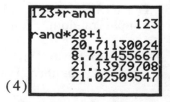

(4)

9. Press **ENTER** three more times for the last two ID numbers of 13 and 24.

> **Note**: rand alone gives a number between 0 and 1, such as the last two lines in screen (5).

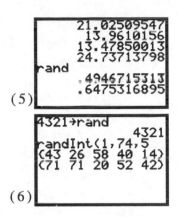

(5)

EXAMPLE: Use the above procedure to pick seven integers from 1 to 74 to play a state Lotto game.

One possibility is 43, 26, 58, 40, 14, 71, and 20, as shown in screen (6).

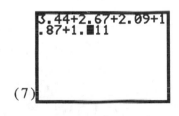

(6)

HOME SCREEN CALCULATIONS AND STORING RESULTS

EXERCISES A [pgs. 27 and 28]: Calculator warmup exercises.

We will use the first two exercises to show some Home screen techniques for dealing with data. Solutions for each of the other exercises (3 to 8) will be given so that you can get some practice with these techniques.

1. $$\frac{3.44 + 2.67 + 2.09 + 1.87 + 3.11}{5}$$

The DEL ↑INS Key, Location 'Ans', and the 'Last Entry Feature'

(a) Type **3.44+2.67+2.09+1.87+1.311**. The last number (1.311) is an intentional mistake.

(b) Press **◄ ◄ ◄** (the left cursor control key three times) to place the cursor over the unwanted 3, as shown in screen (7). Press **DEL**, at C2, and the 3 is deleted. Press **◄ ◄** to move the cursor to the 1 before the decimal, and type **3**, overwriting the 1. Press **ENTER** for the answer of 13.18, the sum of the numerator, as shown in screen (8).

(7)

(c) Press ÷ and 'Ans/' appears on the screen. Type **5** and press **ENTER** for the desired results of Exercise 1 of 2.636.

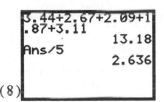

(8)

> **Note**: 'Ans' represents the last value or results of a calculation displayed alone and right justified on the Home screen. Pressing ÷ without typing a value before it called for something to be divided, so 'Ans' was supplied.

(d) Press ↑**ENTRY** to return the 'last entry' or 'Ans/5.' Press ↑**ENTRY** again, and the numerator is returned to the screen.

(e) Press ▲ ▲ (the up cursor control key twice – the first time to move up a line, the second time to jump to the front of the line).

(f) Press ↑**INS**, at B3, to change the blinking rectangular cursor to the blinking underline of the insert mode, as shown in screen (9). Type **(** , the left parenthesis at C6, to insert a left parenthesis before the numerator.

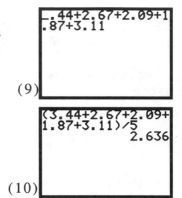

(g) Press ▼ ▼ (the first time to move down a line, the second to jump to the end of the line). Type **)÷5** , to insert a right parenthesis after the numerator and to divide by 5.

(9)

(h) Press **ENTER** for the same results as before (2.636), as shown in screens (10) and (8).

(10)

2.

$$\sqrt{\frac{(2-5)^2 + (4-5)^2 + (9-5)^2}{3-1}}$$

↑**ANS Key, Syntax Errors, and the ALPHA Key and Storage.**

(a) Type **(2-5)²+(4-5)²+(9-5)²** with **2** from the **x²** key at A6. Press **ENTER** for the value 26, as in the top part of the screen (11).

(b) Press ↑**√** (above the **x²** key at A6). On the TI-82, we would have needed to supply a left parenthesis that is part of the TI-83 function. Press ↑**ANS**, at D10; then press **÷(3-1))**, then **ENTER** for the desired results of 3.60555, also shown in the last line of screen (11).

(11)

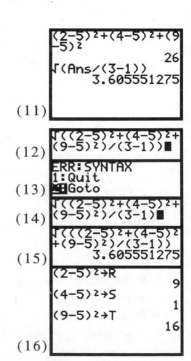

(c) Both (a) and (b) could have been done in one step. Screen (12) shows a first attempt. Pressing **ENTER** brings screen (13), which indicates a syntax error. Press **2** to Goto the error, and the cursor will blink over the last parenthesis as in screen (14). This indicates one more right parenthesis than left, so we need to add another left parenthesis in the beginning, as shown in screen (15).

(12)

(13)

(14)

(15)

(d) We now want to store parts of the above problem in different locations. Type **(2-5)²**, and press **STO►** at A9; then press **ALPHA**, at A3, then **R** at E7, then press **ENTER** for '9' as on top of screen (16). Repeat typing

(16)

$(4-5)^2$ and $(9-5)^2$ and storing in **S** and **T**. Find the sum and store the result in **A** by typing **R+S+T STO► A**. for 26 as shown in the top lines of screen (17) You can then use the sum for the final results by typing **↑√ ALPHA A ÷(3-1))**, then **ENTER** for the last lines of screen (17).

(17)
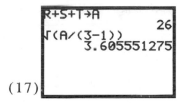

You could do the following exercises differently, but here are some ideas.

3. Note the same answer as Exercise 2.

4. Zero added for balance.

5. It never hurts to be sure with parentheses.

6. Zeros are added for balance.

7. I under **MATH<PRB>4**.

8. Good!

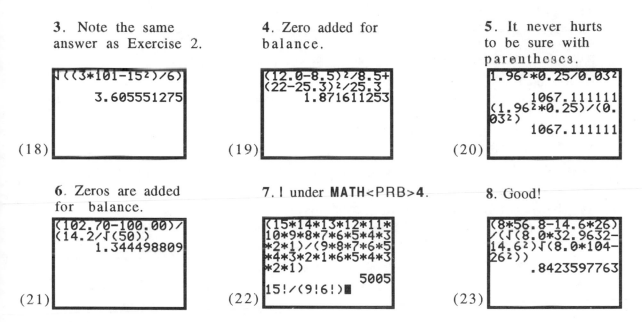

(18)   (19)   (20)

(21)   (22)   (23)

STORING LISTS OF DATA

EXERCISE 9 [pg. 28]: Load and retrieve the data set BLUE (weights of blue M&M® candies (from the data disk). Write the first three values appearing in the first row.

Data can be transferred to your calculator from a data disk, available from the publisher, through a computer with the TI GRAPH LINK system. The data can also be transferred from one calculator to another. See your instructor and/or the appendix for more information.

Storing Lists of Data from the Home Screen

EXERCISE 10 [pg. 28]: Load . . . and save the following tar amounts (in milligrams per cigarette) for 15 different cigarettes. Save the data in list L2 and for the TI-83 in a list named TAR.

16	16	9	8	16	13	15	9	2	15	15	9	14	6	8

1. From the Home screen, engage the left set symbol, {, (found above the left parenthesis). Type in the data above, separating each value with a comma at B6. Press **STO▸** and engage L2 above the **2** key. Press **ENTER** for the set to be repeated without the commas, as in screen (24).

   Note: Use the ▶ key to reveal the values not showing.

   (24)

2. If you made an error in the list as you entered the data, you can correct it on the Home screen. Use the 'last entry' feature to return the list separated by commas, and use the **DEL↑INS** key as in the last section.

   Note: We usually enter data from the STAT editor or spreadsheet, as covered in the next section, because it makes data easier to deal with.

3. To change from milligrams to grams, divide each value by 1000 and store the results in L3, as shown in screen (25) with L2 ÷ **1000 STO▸** L3 **ENTER**.

   (25)

   Note: The ▶ key was used to reveal the last values of the list.

The remainder of this section is for the TI-83.

4. Take values in L2 and store them in a list named TAR, as in screen (26), with L2 **STO▸** ↑**A-LOCK** TAR **ENTER**. Then type **TAR ÷ 1000** and press **ENTER** for a single value result of 3.744 instead of the list of values in L3.

   (26)

5. To see what happened in the last step, type **ALPHA T** and press **ENTER** for the result of 16, as in the last two lines of screen (26). Similarly, A gives 26 and R gives 9. These were the values saved in the last section—you might have different values.

   (27)

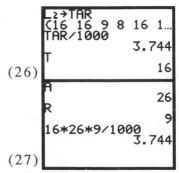

6. We need a way to distinguish between products and list names. Press ↑**LIST** for a screen such as screen (28). Highlight the number, or space, before the list name TAR with the ▼ key, and then press **ENTER** for ʟTAR to be pasted to the Home screen as in screen (29).

   (28)

Note: Under ↑**LIST<NAMES>** the lists are in alphabetical order. If there are numerous list names, TAR might not be on the first screen. Pressing **ALPHA**, and then T advances us to the correct screen. (Pressing ▲ when at the top Name sends us to the last Name.)

(29)

We did not need the small L when we stored a list of data because it was understood that this must be stored to a list. We cannot type TAR alone on the screen, however, if we mean to designate a list. In other situations where it is not clear whether a small L is needed, your safest bet is to paste from ↑**LIST<NAMES>**. This listing of names also has the advantage of giving us the correct spelling of the list saved.

Note: The small L can also be pasted to the Home screen by pressing ↑**LIST<OPS>** B:L **ENTER**, as shown in the last line of screen (29).

Storing Lists of Data Using the STAT Editor or Spreadsheet

SetUpEditor: The TI-82 has only six lists: L1 to L6. The TI-83 can use named lists in addition to L1 to L6 and is limited only by memory size. To make the TI-83 spreadsheet look like the one from the TI-82, with only L1 to L6 showing, press **STAT 5:SetUpEditor** and then **ENTER** for 'Done' as shown in screen (30).

(30)

We want to put these data into the spreadsheet, but first we need to clear out old data.

1. **Clear   Spreadsheet.**

   (31)

   Press **STAT 4**: ClrList L1,L3 **ENTER** (don't forget the comma) for 'Done' on the Home screen, indicating that L1 and L3 are cleared, as shown in screen (31).

2. **View   or   Edit   Spreadsheet.**

   (32)

   STAT 1:Edit... reveals screen (32) with L1 and L3 cleared and with the Tar data in L2.

3. **Enter   Data   into   Spreadsheet.**

   (a) With the cursor at the first row of L1, type **10** **ENTER**. The cursor moves down one row.

   (b) Type **30**, and the screen will look like screen (33), with 30 in the bottom line. Press **ENTER**, and 30 will be pasted in the second row and the cursor

   (33)

   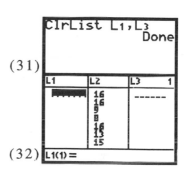

will move down one row.

(c) Continue with 40, 45, 50, 60, and 70.

4. Correct Mistakes with DEL and ↑INS.

(a) In screen (34) we can delete the 45 that we high-lighted by using the ▲ key and then pressing DEL.

(b) To insert a 20 above 30 move the cursor to 30 and press ↑INS to get screen (35), with a 0 where 20 goes.

(c) To insert a 20 type 20 ENTER.

Note: For practice, try deleting a value and then reinserting it.

Note: To check the values of data in a spreadsheet, you can jump a page at a time down or up the list with each press of the green ALPHA key followed by either ▼ or ▲. (The green arrows between these cursor control keys are a visual reminder of this capability.)

L1	L2	L3	1
10	16	------	
30	16		
40	9		
45	8		
50	16		
60	13		
70	15		

(34) L1(4)=45

L1	L2	L3	1
10	16	------	
0	16		
30	9		
40	8		
50	16		
60	13		
70	15		

(35) L1(2)=20■

5. Make a Copy of a List.

Use ▶ and ▲ cursor control keys to highlight L6 as in the top line in screen (36). Engage L2 as in the bottom line of screen (36). Press ENTER and the L2 data appears (not shown).

L4	L5	L6	6
------	------	------	

(36) L6 =L2■

6. Clear Spreadsheet List Without Leaving It.

Use cursor control keys to highlight L2 as in the top line in screen (37). Press CLEAR, then ▼ (or ENTER). Pressing CLEAR reduces the last line to L2= as in screen (37), but the data still exist in L2 until you press ▼ (or ENTER); then they are gone (not shown).

L1	L2	L3	2
10	16	------	
20	16		
30	9		
40	8		
50	16		
60	13		
70	15		

(37) L2 =

Note: I often forget to clear a list before entering the spreadsheet as in **(1)**, so I tend to clear the list this way. Both methods require about the same number of keystrokes, however.

The remainder of this chapter is only for the TI-83.

7. Store Data with a Named List.

Let's store the random numbers 43, 26, 58, 40, 14, 71, and 20 from the Lotto example on page 9 in a new list called RAND2.

(a) First with L2 highlighted, press ↑INS, which moves list L2 to the right and prompts for a Name in the bottom row in ALPHA mode, as in screen (38).

(b) Type RAND and then press ALPHA (to deactivate) and **2**. A flashing quilt pattern will appear indicating that we have used the maximum name length of five characters, as shown in the last line of screen (38).

L1		L2	2
10		------	
20			
30			
40			
50			
60			
70			

(38) Name=RAND2▨

L1	RAND2	L2	2
10		------	
20			
30			
40			
50			
60			
70			

(39) RAND2(1)=43

(c) Press **ENTER** and then ▼ and type **43** for screen (39). Press **ENTER**, and 43 is pasted in the first row. Continue with the other six values.

Note: If the list RAND2 had already been created, its name could have been pasted or typed next to 'Name' in screen (38). Pressing **ENTER** then would not only have pasted the name to the top line but also pasted the data (if it existed) in the rows below the name.

8. Delete a List from the Spreadsheet.

From the spreadsheet, highlight the name on the top line and press **DEL**. The name and the data are gone from the spreadsheet but not from memory.

9. Use **SetUpEditor** to Name List.

(40)

Press and type **STAT 5**: SetUpEditor RAND2,L6,**GRADE**. Then press **ENTER** for 'Done' as in screen (40). Press **STAT 1**:Edit to reveal the new list named GRADE ready for data as in screen (41).

Note: RAND2 can be typed or pasted from ↑**LIST<NAMES>**.

(41)

10. Delete a Named List from Memory.

(a) To remove both the name and data from memory, press ↑**MEM** (above the + key) for screen (42).

(b) Press **2**:Delete... for screen (43).

(42)

(c) Press **4**:List... for a display of list names. Use ▼ to move the selection cursor to the list you want to remove (say GRADE), as shown in screen (44). Be careful!

(43)

(d) Press **ENTER** to delete the list. You can remove lists one by one from this screen.

(e) Press ↑**QUIT** to return to the Home screen.

(44)

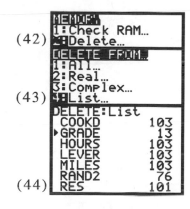

11. For Large List, ↑**LIST<OPS>9:augment** Is Useful.

Several people can cooperate in keying in large data sets. Say part is stored in L1 on one TI-83 and another part is stored in L2 on another TI-83. These data can be shared by linking the two machines and then combining L1 and L2 and storing the results in L3 with ↑**LIST<OPS> 9:augment(L1,L2)STO►L3**. This procedure can be extended to more than two people.

When storing small data sets, using L1 to L6 is convenient. However, it would be wise to name large data sets because you have invested more time to store them.

2

Describing, Exploring, and Comparing Data

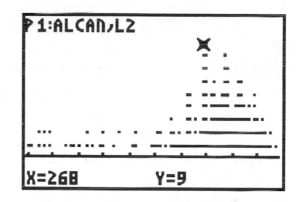

This chapter introduces the plotting and summary statistics capabilities of the TI-83. ↑STAT PLOTS and other keys in the first row are for descriptive plots. Pressing STAT <CALC> 1: 1-Var Stats gives summary statistical calculations. Be sure you have studied Chapter 1 so that you can follow the more abbreviated instructions in this chapter.

FREQUENCY TABLES [pg. 39]

Note: The TI-83 can automate the construction of a frequency table by plotting a histogram from raw data, as explained on page 20.

EXAMPLE [Table 2-2, pg. 39]: The frequency table of the axial loads of 175 aluminum cans is repeated at right. The class marks are included.

Put the class marks in L1 and the frequencies in L2, as explained on page 13, and shown in screen (1).

Class limits Axial Load	L1 Class Mark	L2 Freq.
200-209	204.5	9
210-219	214.5	3
220-229	224.5	5
230-239	234.5	4
240-249	244.5	4
250-259	254.5	14
260-269	264.5	32
270-279	274.5	52
280-289	284.5	38
290-299	294.5	14

Relative Frequency Table [pg. 42]

Relative frequencies can be calculated in L3.

1. Highlight L3 at the top of the column, as in screen (1).

2. Type L2÷175 for the bottom line of screen (1).

3. Press **ENTER** for screen (2). Notice that the first cell (centered at 204.5) has about 5.1% of the data, or 9/175 = .05143, as shown in the first row of L3 (but easier to read from the bottom line of the screen).

   Note: You could have used L2/sum(L2 instead of L2/175 in step 2, with sum pasted from ↑LIST<MATH>.

(1)
```
L1      L2      L3    3
204.5   9       ------
214.5   3
224.5   5
234.5   4
244.5   4
254.5   14
264.5   32
L3 =L2/175█
```

(2)
```
L1      L2      L3    3
204.5   9       .05143
214.5   3       .01714
224.5   5       .02857
234.5   4       .02286
244.5   4       .02286
254.5   14      .08
264.5   32      .18286
L3(1)=.0514285714...
```

Cumulative Frequency Table [pg. 43]

Cumulative frequencies can be calculated in L4.

1. Use the ▶ key to continue beyond L3 to L4. Highlight L4 at the top of the column, as in screen (3).

2. (TI-83 only) Press ↑LIST<OPS>6:cumSum(L2 for the bottom line of screen (3).

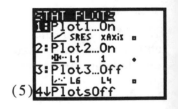

(3)

3. Press **ENTER** for screen (4). Notice that the second row indicates that 9 + 3 = 12 values are in the first two cells, and so on. (Enter these values by hand if you are using a TI-82.)

(4)

EXERCISES 17–20 [pg. 44]: Construct a frequency table using the raw data given in the text.

The TI-83 can automate the construction of a frequency table by plotting a histogram from raw data, as explained on page 20.

HISTOGRAMS FROM FREQUENCY TABLES [pg. 46]

EXAMPLE [pg. 47]: This example continues with the frequency table of axial loads of aluminum cans of the last section with class marks in L1 and frequencies in L2.

1. **Turning OFF All Stat Plots.**
   Activate ↑**STAT PLOT**, in the upper left corner of the keyboard, for a screen such as screen (5). If all plots but Plot1 are not 'Off', press **4:PlotsOff ENTER** for 'Done.'

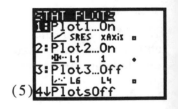

(5)

   Note: You also need to make sure that all plots on the **Y=** edit screen, if you have been using this, are off or cleared. To clear an equation, move the cursor to the right of the equal sign for that function and press **CLEAR**.

2. **Turning ON and Setting Up Stat Plot1.**

   (a) Engage ↑**STAT PLOT** 1: Plot1... for a defining screen which will be set up as on screen (6).

   (b) Use the ▼ ▶ cursor control keys to highlight the choices on screen (6), and press **ENTER** after each choice to activate it. Note that "Type" is a histogram, the third choice in the first row of types, "Xlist" shows the class marks in L1, and the "Frequencies" are in L2.

(6)

      Note: TI-83 users paste the last two list, TI-82 users need to highlight them.

3. Setting Up Plot Windows.

(a) Press the **WINDOW** key (at B1) for a screen with all but the numbers, as screen 7.

(b) At the Xmin= line, type **200** for the lowest class limit from the frequency table.

```
WINDOW
Xmin=200
Xmax=300
Xscl=10
Ymin=-60/4■
Ymax=60
Yscl=0
Xres=1
```
(7)

(c) Press **ENTER** to advance to Xmax=. Enter the lower class limit of the next cell beyond the data, **300**, and then press **ENTER**.

> Note: Class boundaries instead of class limits could have been used in steps (b) and (c) above.

(d) Let Xscl=10 since 10 is the width of the cells (e.g., $210 - 200 = 10$).

(e) Since the maximum frequency is 52, make Ymax=60 and thus Ymin=$^-$60/4, or $^-$15, which is the way this will appear after **ENTER** is pressed in the display screen (7). (The black box cursor is waiting for **ENTER**.)

> Note: The negative symbol is in the bottom row of the keyboard. Set Ymin as the negative of Ymax ÷ 4 to leave room at the bottom of the plot screen for the cell information. Also leave extra room at the top for plot setup information.

(f) Yscl sets the scale of the Y-axis. It was set at 0, so no tick mark will show on the Y, or frequency, axis.

(g) Let Xres=1 or all WINDOWs in this companion.

4. Ploting the Histogram.

Press **TRACE**, wait for the plot to appear, and then press ▶ seven times for the histogram in screen (8). Notice the class limits and the frequency, 52, for the selected cell. To show a graph screen without the added information, press **GRAPH** for the histogram in screen (9).

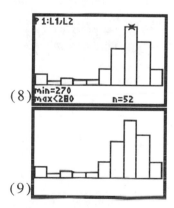
(8)

min=270 max<2■0   n=52

(9)

Relative Frequency Histogram [pg. 46]

EXAMPLE [pg. 47]: This example continues with the frequency table of axial loads of aluminum cans of the last section (and above) with class marks in L1 and relative frequencies in L3.

Adjusting the plot and window setups as shown in screens (10) and (11) gives the results of screen (12),

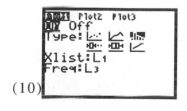
(10)

```
Plot1 Plot2 Plot3
On Off
Type: ⌐ ⌐ ⌐
       ⌐ ⌐ ⌐
Xlist:L1
Freq:L3
```

after pressing **TRACE**, wait for the plot to appear, and then press ▶ seven times for n = 29.7% of the cans having axial loads from 270 to 279 lbs.

Note: Because the TI-82 needs integer values for the frequency list, the relative frequencies would have to be rounded off to the nearest percent in L3 and the WINDOW changed so that Ymin = -9 and Ymax = 36.

(11)

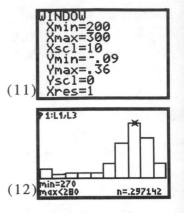

(12)

HISTOGRAMS AND FREQUENCY TABLES FROM RAW DATA [pg. 47]

The first example will be for a data set of 35 values. The second example will continue with the raw data for the 175 aluminum cans.

Note: Because the TI-82 cannot handle lists of sizes larger than 99 values, a special section (the last of this chapter) shows how to handle this larger data set on the TI-82.

EXAMPLE: The survival time of U.S. presidents in years from their inauguration for the 35 presidents from George Washington (10 years) to Lyndon Johnson (9 years) are given here:

10	29	26	28	15	23	17	25	0	20	4	1	34	16	12	4	10	17
16	0	7	24	12	4	18	21	11	2	9	36	12	28	3	16	9	

1. Enter the above data into L6, with numbers entered across by row. (Use the cursor control key ▶ to go to the far right of the spreadsheet for L4, L5, and L6.) Save the data for further use on pages 22 and 29.

2. Under ↑**STAT PLOT** after selecting and turning Plot1 On with 'Type' being the histogram as in screen (13), set
   (a) 'Xlist' at L6 and
   (b) 'Freq' at **1**. This setting means we are going to count each value in L6 one time. There are two 10s in L6, and each one is counted once and put in the proper cell; in other words, the frequency of that cell is increased by 2.

(13)

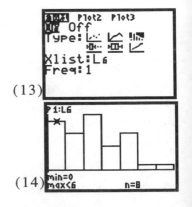

3. **Automatic Histogram** (For TI-82, skip to step 5.) Press **ZOOM 9:ZoomStat** and then **TRACE** for screen (14) with 8 presidents who lived less than 6 years after their inauguration. By using the ▶ key, you can find 6, 9, 4, 6, 1, and 1 in the other cells.

(14)

4. This is a good first look at the data, but you might want to set your own cell limits. Press **WINDOW** for screen (15). (In the next example you will see that cell widths are usually not integers such as the Xscl = 6 of screen (15).)

5. Change the values in the **WINDOW**, and set the cell width, or Xscl = 10, as in screen (16). Since the last cell of the previous histogram went from 36 to <42, we know that the largest data value is 36, so Xmax = 40 is sufficiently large.

6. Press **TRACE** for the histogram in screen (17). By using the ▶ and ◀ keys, even though the histogram does not fit on the screen, we find the four cells have values of 11, 13, 9, and 2.

7. Adjust the **WINDOW** height by letting Ymin= ‾4 and Ymax=16, as in screen (18).

8. Press **TRACE** again for the histogram in screen (19) followed by its frequency table.

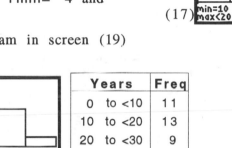

```
(15) WINDOW
     Xmin=0
     Xmax=42
     Xscl=6
     Ymin=-2.70621
     Ymax=10.53
     Yscl=0
     Xres=1
```

```
(16) WINDOW
     Xmin=0
     Xmax=40
     Xscl=10
     Ymin=-2.70621
     Ymax=10.53
     Yscl=0
     Xres=1
```

(17) P 1:L6   min=10 max<20   n=13

```
(18) WINDOW
     Xmin=0
     Xmax=40
     Xscl=10
     Ymin=-4
     Ymax=16
     Yscl=0
     Xres=1
```

(19) P 1:L6   min=10 max<20   n=13

Years	Freq
0 to <10	11
10 to <20	13
20 to <30	9
30 to <40	2

EXAMPLE [pg. 47]: The axial load capability of 175 aluminum cans [Table 2-1] is repeated below and is to be stored in list LALCAN, with numbers entered across by row-order. (This type of ordering is important for Chapter 12.)

270	273	258	204	254	228	282	278	201	264	265	223	274	230	250	275	281
271	263	277	275	278	260	262	273	274	286	236	290	286	278	283	262	277
295	274	272	265	275	263	251	289	242	284	241	276	200	278	283	269	282
267	282	272	277	261	257	278	295	270	268	286	262	272	268	283	256	206
277	252	265	263	281	268	280	289	283	263	273	209	259	287	269	277	234
282	276	272	257	267	204	270	285	273	269	284	276	286	273	289	263	270
279	206	270	270	268	218	251	252	284	278	277	208	271	208	280	269	270
294	292	289	290	215	284	283	279	275	223	220	281	268	272	268	279	217
259	291	291	281	230	276	225	282	276	289	288	268	242	283	277	285	293
248	278	285	292	282	287	277	266	268	273	270	256	297	280	256	262	268
262	293	290	274	292												

Set up **Plot1** as in screen (20). Press **ZOOM 9:ZoomStat** and then **TRACE** with a few ▶s for the histogram in screen (21), with the cell of maximum frequency having 60 values. Press **WINDOW** to reveal the cell width of Xscl = 12.125, as in screen (22).

(20)

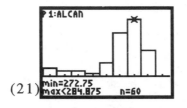

(21)

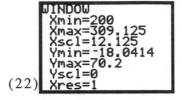

(22)

Adjust the WINDOW values as in screen (7), page 19, and press **TRACE** to reveal the same histogram as screen (8) with the cell counts of the frequency table on page 17.

DOT PLOTS [pg. 48]

EXAMPLE: The survival time data of U.S. presidents in years from their inauguration for the 35 presidents from George Washington to Lyndon Johnson is repeated below from page 20. Hopefully you saved them in L6 (if not, reenter them there now).

10	29	26	28	15	23	17	25	0	20	4	1	34	16	12	4	10	17
16	0	7	24	12	4	18	21	11	2	9	36	12	28	3	16	9	

A dot plot is not difficult to do by hand if the data is put in order with the TI-83, but you can construct the entire dot plot on the TI-83.

1. **Sorting Data in Ascending Order.**
   (a) The presidents data is currently listed in L6 in the order of George Washington, the 1st president, to Lyndon Johnson, the 35th president. Make a copy of this data by storing it in L1 from the Home screen, with L6 **STO►** L1, or alternately by highlighting L1 in the spreadsheet and pasting L6 in the bottom line of the screen for L1 = L6 and then pressing **ENTER**.
   (b) Press **STAT 2**: SortA(L1 and then **ENTER** for 'Done' as in screen (23). The data is now in ascending order (in L1) from low value to high as revealed by screen (24) of the **STAT 1**:Editor.

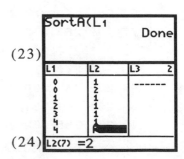

(23)

(24)

2. Count the frequency that each data value occurs. In L2, type the counting numbers next to each year value in L1. There are 1, 2 values of 0 years survived after inauguration. Only one president survived 1 or 2 or 3 years after inauguration. (John F. Kennedy was assassinated after 3 years as president. There are 1, 2, 3

presidents who survived 4 years. The complete list is given in the table at the far right.

3. Set up Plot1 for the first 'Type' (a scatter plot) as in screen (25).

4. Press **ZOOM 9:ZoomStat** and then **TRACE** for the first effort in the plot of screen (26). (Set **AxesOff** under ↑**FORMAT** (at C1), if you wish, but do not forget to return to **AxesOn** later.)

5. Set the **WINDOW** as in screen (27) with Ymin and Ymax values that work well for moderate-size data sets. Press **TRACE** with a few ▶s for screen (28), with the highlighted point indicating three presidents survived 4 years.

Note: Make sure your two lists (L1 and L2) are the the same length or you will get a Dimension Mismatch Error.

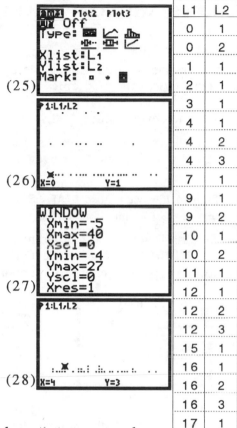

(25)

(26)

(27)

(28)

L1	L2
0	1
0	2
1	1
2	1
3	1
4	1
4	2
4	3
7	1
9	1
9	2
10	1
10	2
11	1
12	1
12	2
12	3
15	1
16	1
16	2
16	3
17	1
17	2
18	1
20	1
21	1
23	1
24	1
25	1
26	1
28	1
28	2
29	1
34	1
36	1

EXAMPLE [pg. 48]: The 175 aluminum can data values that are stored in list LALCAN were plotted in the screen next to this chapter's title (page 17). The screen will look better than this on your TI-83; some points are too close together for the printer to distinguish.

Note: I took a little poetic license in using LALCAN for the sorted data in this example. The rest of this chapter has the data in the original order.

STEM-AND-LEAF PLOTS [pg. 49]

EXAMPLE: The survival time data of U.S. presidents is sorted as in the above example and stored in L1 as shown in the table at the far right. It is easy to build the stem-and-leaf plot from this sorted data at the near right.

0	00123444
0	799
1	001222
1	5666770
2	0134
2	56889
3	4
3	6

PARETO AND BAR CHARTS [pg. 51]

EXAMPLE [pg. 51]: In a recent year, 75,200 accidental deaths in the United States were attributable to the accident types in the table at the right.

Accident Type	Deaths
Motor vehicles	43500
Falls	12200
Poison	6400
Drowning	4600
Fire	4200
Ingestion	2900
Firearms	1400

1. Put the seven values 1, 2, 3, 4, 5, 6, 7, in L1.

2. Put the number of deaths for the seven types into L2 as shown in screen (29).

   Note: If the data were not in descending order from high to low value, use **STAT 3:SortD(L2** (similar to what was done for dot plots on page 22).

(29)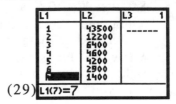

3. Set up Plot1 as a histogram in screen (30), and the **WINDOW** as in screen (31), and press **TRACE** for the pareto chart in screen (32).

(30) 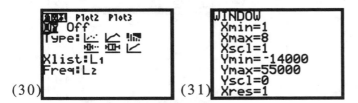 (31)

(32)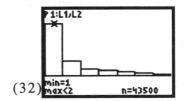

4. A bar chart of the above data has a gap between the different categories. [See Figure 1-1(a) on pg. 13]. This gap is obtained by changing Xscl = 0.5 in the WINDOW of screen (31) and then pressing **TRACE** for screen (33).

(33)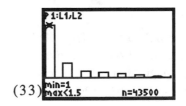

PIE CHARTS [pg. 51]

EXAMPLE: The TI-83 **STAT 1:**Editor can help us make a pie chart of the 75,200 accidental deaths from the example above by calculating the percentages for each type and the degrees in the central angles.

1. With the data in L2, highlight L3, as in screen (34). Type **100L2÷sum(L2** as in the bottom line of the same screen, with 'sum(' from ↑LIST<MATH>5:sum(.

2. Press **ENTER** for the percentages of each type of accident in L3, as in screen (35). Highlight L4, in the top line, and type **360L2÷sum(L2** as in the bottom line of the same screen.

(34)(35)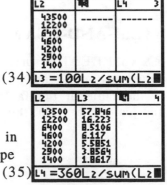

3. Press **ENTER** for the center angle in degrees of each type of accident in L4, as shown in screen (36).

To summarize, in the first row of screen (36), the 43500 deaths due to motor vehicles represent 57.8% of the deaths, or 208 degrees of the pie.

(36)

SCATTER DIAGRAMS [pg. 52]

EXAMPLE [Data Set 4, Appendix B]: The cigarette nicotine and tar data is given in the table at the right. Plot the scatter diagram of the data pairs, with the nicotine values for the horizontal (X) axis and the tar values for the vertical (Y) axis.

1. With the X values in L1 and the y values in L2 set up Plot1 for the first 'Type' as shown in screen (37). All other plots, including Y= plots, must be off.

(37)

2. Press **ZOOM** 9:ZoomStat and then **TRACE** for the plot of screen (38).

3. For the **regression line**, press **STAT<CALC>8:LinReg(a+bx) L1,L2,Y1** for screen (39) with Y1 from **VARS<Y-VARS>1:Function 1:Y1**. (The TI-82 uses **9** instead of **8** and cannot use Y1.)

(38)

4. Press **ENTER** for screen (40). Your screen might look different and contain more information. This variation will be covered in Chapter 9.

(39)

(40)

5. Press **ZOOM** 9:ZoomStat and then **TRACE** for the plot with the regression line of screen (41). (The TI-82 must enter the equation of the regression line into the Y= editor before this step, with Y1 = 14.21x - 1.27.)

(41)

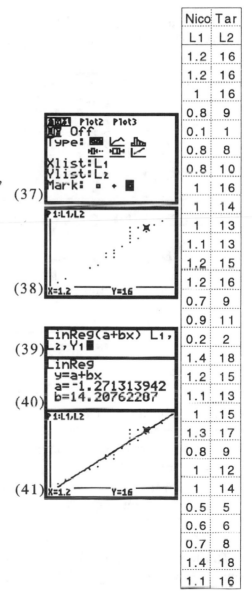

Nico	Tar
L1	L2
1.2	16
1.2	16
1	16
0.8	9
0.1	1
0.8	8
0.8	10
1	16
1	14
1	13
1.1	13
1.2	15
1.2	16
0.7	9
0.9	11
0.2	2
1.4	18
1.2	15
1.1	13
1	15
1.3	17
0.8	9
1	12
1	14
0.5	5
0.6	6
0.7	8
1.4	18
1.1	16

FREQUENCY POLYGONS [pg. 58]

EXERCISE 27 [pg. 58]: Construct a frequency polygon from the aluminum can data from page 17 repeated in the table below with an extra class mark on each side of the data given a frequency of zero.

ClassMark(L1)	194.5	204.5	214.5	224.5	234.5	244.5	254.5	264.5	274.5	284.5	294.5	304.5
Frequency(L2)	0	9	3	5	4	4	14	32	52	38	14	0

With the data in L1 and L2, set up Plot1 for an xyLine plot as in screen (42). Press **ZOOM** 9:ZoomStat and then **TRACE** for the frequency polygon plot in screen (43).

(42)

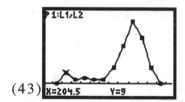
(43)

MEASURES OF CENTRAL TENDENCY WITH 1-Var Stats [pg. 59]

EXAMPLE [pg. 61]: Listed below are the times (in years) that the first ten presidents survived after inauguration. Find the mean and median for this sample.

10 29 26 28 15 23 17 25 0 20

Mean and Median from Raw Data

1. Put the data in L1.
   Press **STAT** <CALC> **1**: 1-Var Stats **L1** for screen (45).

   Note: Pressing **STAT** shows the EDIT menu, ▶ shows the CALC menu, as in screen (44), and **1** pastes "1-Var Stats" to the Home screen. Then engage **L1** for screen (45).

2. Press **ENTER** for the first of the two screens of output, screen (46). (We will call these the first and second screens of output.) To reveal the second screen of output, screen (47), hold down the ▼ cursor control key.

   The mean = $\bar{x} = \Sigma x / n$ = 193/10 = 19.3 on the first screen of output, and the median = Med = 21.5 on the second screen of

(44)
```
EDIT CALC TESTS
1:1-Var Stats
2:2-Var Stats
3:Med-Med
4:LinReg(ax+b)
5:QuadReg
6:CubicReg
7↓QuartReg
```

(45)
```
1-Var Stats L1
```

(46)
```
1-Var Stats
 x̄=19.3
 Σx=193
 Σx²=4469
 Sx=9.09273214
 σx=8.626123115
↓n=10
```

(47)
```
1-Var Stats
↑n=10
 minX=0
 Q1=15
 Med=21.5
 Q3=26
 maxX=29
```

output. We will discuss more of the 1-Var Stats output later.

Midrange [pg. 64]

The midrange is easily calculated from the values in
screen (47) with $(maxX + minX) \div 2 = (29 + 0) \div 2 = 14.5$.

(48)

(49)

EXAMPLE [pg. 64]: Find the mean, median, and
midrange for the 175 axial loads of aluminum cans
saved in list LALCAN.

Press **STAT <CALC> 1**: 1-Var Stats LALCAN and then **ENTER**
for screens (48), (49), and (50) with the mean = $\bar{x} = 267.1$,
the median = Med = 273, and the midrange = $(297 + 200) \div 2 =$
248.5.

(50)

Mean from a Frequency Table [pg. 66]

EXAMPLE [pg. 67]: The frequency table of the axial
loads of 175 aluminum cans is repeated from page 17
in the table at the right with class marks stored in L1
and frequencies in L2.

(51)

Press **STAT <CALC> 1**: 1-Var Stats **L1,L2**
and then **ENTER** for the output in
screen (52), with mean = $\bar{x}$ = 267.1857
= 267.2 compared to the exact value from
the raw data of 267.1 in screen (49).

(52)

| L1 | L2 |
Class Mark	Freq.
204.5	9
214.5	3
224.5	5
234.5	4
244.5	4
254.5	14
264.5	32
274.5	52
284.5	38
294.5	14

Weighted Mean [pg. 66]

EXAMPLE [pg. 66]: Find the mean of five test scores
(85, 90, 75, 80, 95) if the first four tests count for 15%
each and the final score counts for 40%.

(53)

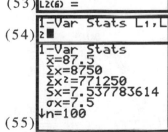

(54)

Put the scores in L1 and the weights in L2, as in screen
(53). Press **STAT <CALC> 1**: 1-Var Stats **L1,L2** and then **ENTER**
for the output in screen (55) with the weighted mean =
$\bar{x}$ = 87.5.

(55)

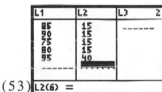

MEASURES OF VARIATION [pg. 74]

EXAMPLE [pg. 77]: The waiting times (in minutes) for a sample of ten customers at the Jefferson Valley Bank are 6.5, 6.6, 6.7, 6.8, 7.1, 7.3, 7.4, 7.7, 7.7, 7.7.

1. **Standard Deviation and Variance.**
   With the data in L1, press
   STAT [CALC] **1**:1-Var Stats L1 **ENTER**. This sequence
   gives screens (57) and (58), with the
   standard deviation of Sx=0.4766783215 min.

   Variance = Sx^2 = 0.4766783215^2 = 0.2272222222 min^2.

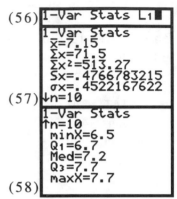

(56)

(57)

2. **Range** = maxX - minX = 7.7 - 6.5 = 1.2 minutes, which is
   easy enough to see with the data in order or a small
   data set but difficult with unordered data and larger
   data sets.

(58)

3. **VARS 5:Statistics Menu.**

   (a) Press **VARS** at D4 for the VARS menu of screen (59).
   (b) Press **5** for the Statistics submenus of screen (60).
   (c) Press **3**, and Sx is pasted to the Home screen.

   Note: In short form; **VARS 5**:Statistics **3**:Sx.

   (d) Press **ENTER** for 0.4766783215, as in screen (61)
   and Step 1 above.

   (e) Press the x^2 key at A6 for Ans^2 (or Sx^2) and then
   ENTER for .2272222222 as before.

   Note: This saves typing in all the digits of Sx or rounding
   it off, but pasting Sx makes sense only after you have
   performed 1-Var Stats or another appropriate calculation;
   otherwise, some past calculation from other data will be
   stored in Sx.

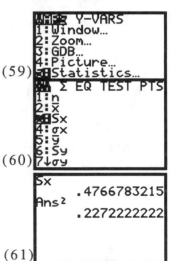

(59)

(60)

(61)

Standard Deviation from a Frequency Table [pg. 81]

EXAMPLE: The frequency table of the axial loads of
175 aluminum cans with class marks stored in L1 and
frequencies in L2 used **STAT** <CALC> **1**: 1-Var Stats **L1,L2**
and then **ENTER** for the output of screen (63), which is
a repeat of screen (52).

Sx = 22.26 compared to the exact value from the raw data
of 22.11, as in screen (49).

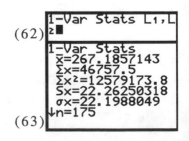

(62)

(63)

MEASURES OF POSITION

Quartiles, Deciles, and Percentiles [pg. 94]

EXAMPLE: Let's return to the presidents data and sort it in ascending order in list L1 as was done on page 22. We will also use list LALCAN of 175 axial loads of aluminum cans and sort them in ascending order in list L2.

1. [pg. 95] Find the percentile corresponding to the president who survived 17 years after inauguration and of the can that withstood an axial load of 241 pounds.

   (a) Press **STAT** 1:Edit and use the ▼ cursor control key to go to 17 in L1, which just happens to be the same row as 241 in L2 (how convenient!). This is the 22 row, as shown in the notation in the bottom line of screen (64), so 21 values are less.

   (b) This is the 60th percentile of presidential survival times and the 12th percentile of can strength as calculated in screen (65).

L1	L2	L3	1
12	225		
12	228		
15	230		
16	230		
16	234		
16	236		
	241		

(64) L1(22) =17

(65)
```
(21/35)*100
                 60
(21/175)*100
                 12
```

2. [pg. 96] Find the 25th percentile, or P_{25}.

   (25/100)*35 = 8.75, so we take the 9 value, or **L1(9)**, which is seven years, as shown in screen (66). **L1(9)** is obtained by pasting L1 to the Home screen and then using the parentheses keys.

   Similarly, the 25th percentile of can strength is 262 pounds, as shown in the bottom lines of screens (66) and (67).

(66)
```
(25/100)*35
              8.75
L1(9)
                 7
(25/100)*175
             43.75
```

(67)
```
L2(44)
               262
```

   Since $P_{25} = Q_1$, we would hope that the TI-83 value for the first quartile is close to what we calculated above, and it usually is, although we calculated it by another method. For this example, they agree exactly as the second output screens, (69) and (70), from **STAT <CALC> 1**: 1-Var Stats.

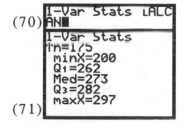

(68)
```
1-Var Stats L1█
```

(69)
```
1-Var Stats
↑n=35
 minX=0
 Q1=7
 Med=15
 Q3=23
 maxX=36
```

(70)
```
1-Var Stats LALC
AN█
```

(71)
```
1-Var Stats
↑n=175
 minX=200
 Q1=262
 Med=273
 Q3=282
 maxX=297
```

3. [pg. 98] Find the 40th percentile, or P_{40}, for the axial load data for aluminum cans.

Since 40% of 175 is 70, an integer value, take the average of the 70th and 71st value for $P_{40} = 269$, as shown in screen (72).

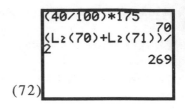

(72)

BOXPLOTS AND FIVE-NUMBER SUMMARY [pg. 104]

EXAMPLE [pg. 104]: Use the pulse rates of the following sample of smokers to (a) find the values of the five-number summary and (b) construct a boxplot for the pulse rates of these smokers. (First store the data in L1.)

| 52 | 52 | 60 | 60 | 60 | 60 | 63 | 63 | 66 | 67 | 68 |
| 69 | 71 | 72 | 73 | 75 | 78 | 80 | 82 | 83 | 88 | 90 |

Five-Number Summary

Press STAT <CALC> 1: 1-Var Stats L1 and then ENTER, and use the ▼ key for the second screen of output of screen (74) with minX, Q_1, Med, Q_3, maxX, or 52, 60, 68.5, 78, 90, the **five-number summary**.

(73)

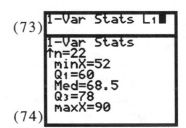

(74)

Boxplot (or Box-and-Whisker Diagrams)

1. Using ↑STAT PLOT, set up Plot1 as in screen (75). Under 'Type,' the fourth option (or the first in the second row), is highlighted. (WINDOW has Xscl = 10.)

2. Press ZOOM 9:ZoomStat and then TRACE, which will automatically show the plot of screen (76). (When you use this procedure on other problems, if one or more points separate themselves from the whiskers, you have a possible **outlier**—read the next section for details.)

(75)

3. You can use the ▶◀ keys to move right and left within a boxplot and to display the five-number summary values below the plots, as shown in the last line of screen (76) with Med(ian) = 68.5.

(76)

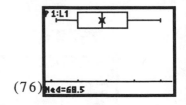

OUTLIERS AND SIDE-BY-SIDE PLOTS [pgs. 105, 107]

EXERCISES 3 AND 12 [pgs 107, 108]: Use side-by-side boxplots to compare the ages of actors and actresses at the time they won Oscar awards.

Actors (L1):	32	37	36	32	51	53	33	61	35	45	55	39
	76	37	42	40	32	60	38	56	48	48	40	
	43	62	43	42	44	41	56	39	46	31	47	
Actresses(L2):	50	44	35	80	26	28	41	21	61	38	49	33
	74	30	33	41	31	35	41	42	37	26	34	
	34	35	26	61	60	34	24	30	37	31	27	

1. Store the actors ages in L1 and the actresses ages in L2.

2. In addition, make a copy of the actresses ages in L3, and then delete the one age of **60**.

3. Using **↑STAT PLOT**, set up Plot1, Plot2, and Plot3 with settings as in screen (77). (**WINDOW** has Xscl = 10.)

4. Press **ZOOM** 9:ZoomStat, then **TRACE**, then ▼ ▼, and then ▶ ▶ for screen (78).

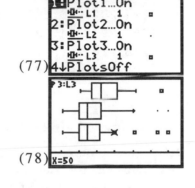

(77) (78)

Note: The extra points to the right are possible outliers, as explained in Exercise 12, and do not show up with the TI-82, which extends the whiskers out to the maximum values to the right. (These possible outliers can also be hidden with the TI-83 by using the fifth 'Type' option for a regular boxplot; we have been using the modified boxplot because it gives you more information.) The TI-83 does not distinquish between mild and extreme outliers. Screen (80) shows another possibility that also works on the TI-82.

5. Use ▼ to go from plot to plot. You can use the ▶ ◀ keys to move right and left within a boxplot. The point showing in the bottom line and bottom plot of screen (78) is an actress who is 50 years old. We deleted the next age (60) from Plot2 because it was so close to the 61 (actually two 61s as revealed when traced) that shows clearly in Plot3.

   Note: Twenty-five percent of the men are above Q3 = 51 while 5 of 34 = 15% of the women were that old when they received Oscars. Fifty percent of the men were younger than the median of 42.5 years while 75% of the women were younger than 42 years.

6. **Boxplot with Histogram**
   Turn off Plot1, change Plot3 to a histogram for the data in L2, and change **WINDOW** as in screen (79). Press **TRACE**, ▼, and a few ▶s for screen (80). Notice that the two values at 61 years show up at the highlighted point.

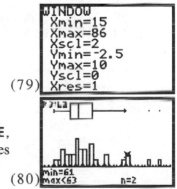

(79) (80)

TI-82 WITH A LIST OF MORE THAN 99 VALUES (OPTIONAL)

EXAMPLE [Table 2-1]: The axial load capability of 175 aluminum cans is repeated here. Put the first 99 values into L5 and the last 76 values in L6.

270	273	258	204	254	228	282	278	201	264	265	223	274	230	250	275	281
271	263	277	275	278	260	262	273	274	286	236	290	286	278	283	262	277
295	274	272	265	275	263	251	289	242	284	241	276	200	278	283	269	282
267	282	272	277	261	257	278	295	270	268	286	262	272	268	283	256	206
277	252	265	263	281	268	280	289	283	263	273	209	259	287	269	277	234
282	276	272	257	267	204	270	285	273	269	284	276	286	273	289	263	270
279	206	270	270	268	218	251	252	284	278	277	208	271	208	280	269	270
294	292	289	290	215	284	283	279	275	223	220	281	268	272	268	279	217
259	291	291	281	230	276	225	282	276	289	288	268	242	283	277	285	293
248	278	285	292	282	287	277	266	268	273	270	256	297	280	256	262	268
262	293	290	274	292												

Frequency Table from Histograms

Define the plots as in screens (81) and (82). Set the **WINDOW** as in screen (83) and press **TRACE** for the two histograms superimposed as shown in screen (84). Turn on one plot at a time for separate plots as shown in screens (85) and (86).

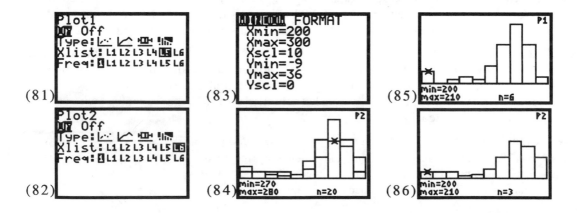

Sum the values in each interval. For example 6 + 3 = 9 for the first interval. The sum for the ten intervals are 9, 3, 5, 4, 4, 14, 32, 52, 38, 14, from which we can make a frequency table that agrees with the values on page 17.

Means   and   Standard   Deviations

Use 1-Var Stats on L5 and L6 for screens (88) and (90).
The mean for all values equals the total for each list
(Σx) divided by the total number of values in each list.

Mean = (26343 + 20402)/(99 + 76) = 267.1142857,
which agrees with screen (49) on page 27.

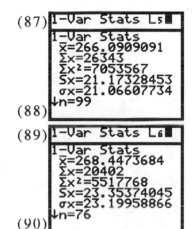

(87)

(88)

(89)

(90)

The variance for the 175 values can be obtained by
working with Σx^2, Σx, and n of both groups, as follows.

(7053567 + 5517768 − (26343 + 20402)^2/(99 + 76))/(99+76-1)
for s^2 = 488.952381. Therefore, s = $\sqrt{}$(488.952381) =
22.11226766.

3

Probability

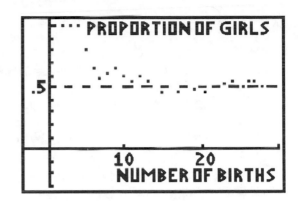

In this chapter you will learn some helpful techniques for calculating probabilities, such as using the change-to-fraction function and the raise-to-the-power key, and for calculating factorials, permutations, and combinations. You will also be introduced to simulations, which not only give approximate values for probabilities but also help elucidate probability concepts.

THE LAW OF LARGE NUMBERS [pg. 124]

As an experiment is repeated again and again, the relative frequency probability of an event tends to approach the actual probability.

EXAMPLE [Figure 3-2, pg. 125]: The figure above shows computer-simulated results of the law of large numbers. Notice that when the number of births is small, the proportion of girls greatly fluctuates. As the number of births increases, the proportion of girls approaches 50%. If we flip coins and let H, or a head, represent a girl birth, and T, or tail, represent a boy birth, we can model this situation by flipping many coins. But it is much easier to let the computer do the simulation.

The following steps for doing this simulation show some of the possibilities for doing simulations with your calculator. Follow along with your TI-83. (TI-82 users should read along in order to understand the TI-82 results in step 8.)

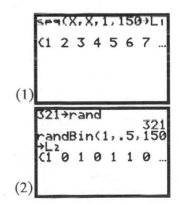

1. Put the numbers 1 to 150 in L1 with
   ↑LIST<OPS>5:seq(X,X,1,150 STO► L1, as in the screen (1).

2. Use 321 STO► rand ENTER to set the seed (as explained on page 8) so that you can follow along. (See the first two lines of screen (2).)

3. Generate one (1) random coin toss at a time with a 0.5 chance of getting a Girl = G = 1. Do this 150 times and store the results in L2 with
   MATH<PRB>7:randBin(1,.5,150 STO► L2. The results start with {1 0 1 0 1 1 0 ... and can be read as {G B G B G G B ... .

4. ↑LIST<OPS>6:cumSum(L2 STO► L3 ENTER stores the
   cumulative sum of L2 in L3. (See screen (3).). The results
   {1 1 2 2 3 4 4 ... indicate that the first girl occurred on the
   first birth (1), the second girl on the third birth (1 + 0 +
   1 = 2), the third girl at the fifth birth (1 + 0 + 1 + 0 + 1 = 3),
   and so on.

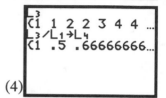
(3)

5. L3÷L1 STO►L4 gives the proportion of heads, or {1 .5
   .6666... , indicating 100% (1/1=1) girls on the first toss,
   50% (or 1 of 2 heads after the second toss since there is
   still just one head), and 66.67% heads (or 2 out of 3 heads)
   after the third toss. This is a lot of fluctuation in the
   short run.

(4)

6. To see what happens in the long run, set up and plot
   the results as follows.
   (a) Press ↑STAT PLOT 1:Plot1 to set up for an xyLine plot
       as in screen (5).
   (b) Press the Y= key, and set Y1 = .5 to plot this
       horizontal line as in screen (6).

(5)

   (c) Set WINDOW as shown in screen (7) and then press
       TRACE for the screen (8) plot of the first 20 tosses
       since Xmax = 20.
   (d) Change Xmax = 100, leaving the other values as before.
       (See screen (9).) Press TRACE for the plot of screen
       (10), which covers 100 tosses. The first tick mark on the
       X-axis marks off the first 10 tosses, and the plot to that
       point is the same as in screen (8) but condensed by the
       change of scale. Notice how the plotted points hover
       about the Y1 = 0.5 line.

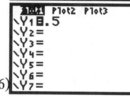
(6)

(7)   (8)   (9)   (10)

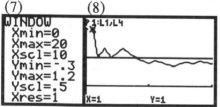   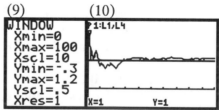

7. You can check intermediate results as shown in screen
   (11) (as explained on page 29) or quickly calculate a
   proportion after a given number of tosses, for example,
   150, with randBin(150,0.5)÷150 as in screen (12).

(11)

8. The last results can be duplicated on the TI-82 as in
   screens (13) and (14) for the proportion after 99 and
   150 tosses. (iPart is from MATH<NUM>, and mean and sum
   are from the ↑LIST<MATH> menus.) All the other steps

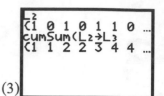
(12)

could also be duplicated (up to the 99 value limit for a list on the TI-82) by making similar substitutions.

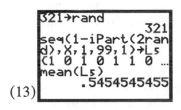

(13)

(14)

TWO USEFUL FUNCTIONS

Change to Fraction Function

EXERCISE 13b [pg. 131]: If a person is randomly selected, find the probability that his or her birthday is in November. Ignore leap years.

1. Since November has 30 days and a year has 365 days, each of the 365 days are equally likely for a randomly selected individual. The probability is 30 ÷ 365 = 0.082. (See the first two lines in screen (15).)

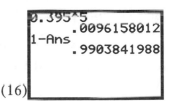

(15)

2. Press **MATH 1:▶Frac**, which gives Ans ▶Frac, and then **ENTER**, which gives 6/73 (or 30/365 reduced), as in screen (15).
   Note: This change-to-fraction function is handy when you prefer to give the answer as a fraction.

The Raise-To-the-Power Key
Probability of "At Least One" [pg. 147]

EXAMPLE [pg. 147]: Find the probability that at least one of five fellow employees in San Francisco has a listed telephone number (and can therefore be called). Assume that telephone numbers are independent and that for San Francisco, 39.5% of the numbers are unlisted.

1. P(5 unlisted numbers among five employees) =
   $0.395*0.395*0.395*0.395*0.395 = 0.395^5$

   Type **0.395^5** and then press **ENTER** for 0.00962, as in the first two lines of screen (16).
   Note: Use the raise-to-the-power key ^ at E5.

2. Type **1 - ↑ANS** and then press **ENTER** for 0.99038, the complement of the above and the solution to the problem. (See the last lines of screen (16)).
   Note: ↑ANS is at D10.

(16)

PROBABILITIES THROUGH SIMULATION [pg. 155]

Finding probabilities of events can sometimes be difficult. We can often benefit from using simulation. You should review the first example in this chapter before doing the examples in this section. Each example assumes you are familiar with the examples that came before it.

EXAMPLE [pg. 155]: In testing techniques of gender selection, medical researchers need to know probabilities related to the gender of offspring. Assuming that males and females are equally likely, describe an experiment that simulates gender from births.

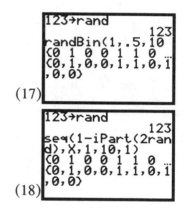

One simulation is simply to flip a coin, with heads representing male and tails representing female. Another approach is to use the TI-83 to randomly generate 0s and 1s, with 0 representing male and 1 representing female. This technique was used on the first page of this chapter, but we will duplicate this technique for 10 births as shown in screens (17) and (18). Screen (17) is only for the TI-83, whereas the technique in screen (18) also works for the TI-82.

(17)

Note: The complete list of 10 values was shown by pressing ↑**RCL** (at A9), then ↑**Ans**, and then **ENTER**.

Based on the results, this family had one boy followed by (18) a girl but did not have an equal number of boys and girls again until after the sixth birth.

EXAMPLE [pg. 156]: One classic exercise in probability is the <u>birthday problem</u>, in which we find the probability that in a class of 25 students, at least 2 students have the same birthday. Ignoring leap years, describe a simulation of the experiment that yield birthdays of 25 students in a class.

First set the seed as in the first two lines of screen (19). Generate 25 random integers between 1 and 365, and store these results in L1. Put the data in order, and then display the results on the Home screen with
MATH<PRB> 5:randInt(1,365,25 STO► L1
ALPHA : STAT 2:SortA(L1 ALPHA : L1. (19)
Press **ENTER** and then use **►** to view the values in the sorted list. You will discover two 190s, or two students who have the same birthday. (See the last line of screen (19).)

Note: The colon (above the decimal key at C10) lets us keep several statements together.

Press **ENTER** again and check the birthdays of the next class of 25 students (not shown in screen (19) but shown in screen (20), which is a version that works on both the TI-83 and TI-82). Checking the second list reveals that no pairs have the same birthday.

(20)

Just with these two tries, we have a 50% chance of getting at least two students in the class with the same birthday. But two tries are not enough. Keep pressing **ENTER** and investigating the resultant lists for more ties and for a better estimate of the desired probability.

EXAMPLE [pg. 156]: The Delmarva Communications Company manufactures cellular phones and has been experiencing a 6% rate of defects. The quality control manager knows that the phones are produced in batches of 250. On average, then each batch contains 15 defects (0.06*250 = 15). The quality control manager wants to know how much the number of defects will typically vary. Describe a simulation of 250 cellular phones that are manufactured with a 6% rate of defects.

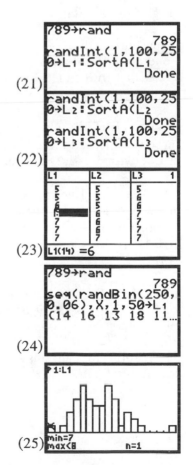

(21)

After setting a random seed, generate 250 integers between 1 and 100, stored in L1 and then put in order as in screen (21). (This is similar to what we did in the previous example.) Also, generate samples for L2 and L3, as in screen (22). Integers 1 to 6 represent defective phones, and integers 7 to 100 represent nondefective phones.

(22)

Press **STAT 1:Edit** and ▼ to reveal screen (23). There are 14 values of the integers 6 or less (representing defects) in L1, as shown in the last line of screen (23). Likewise, there are 16 defects in L2 and 13 defects in L3.

(23)

You can duplicate and extend this simulation to 50 samples, as in screen (24). Notice that the first three values stored in L1 and shown in the last line of screen (24) are 14, 16, and 13 defects, as before. (The output takes a bit of time to calculate, so be patient.)

(24)

A histogram of the 50 samples in L1 shows that the number of defects varies. The low is 7 (see screen (25)), and most are less than 22, except for one sample of 26 defects, which is at the far right in screen (25). **Note**: Press **ZOOM 9:ZoomStat** and then change Xscl = 1 and press **TRACE** for a similar screen with Plot1 set for a histogram.

(25)

Screens (26) and (27) show a program (in the program editor) and its output that work for both the TI-83 and the TI-82 and are easy to modify (the first line of the program) to simulate more than three samples.

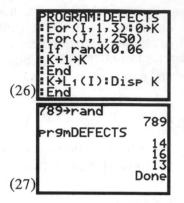

(26)

(27)

EXAMPLE [pg. 157]: Simulate rolling a pair of dice 99 times. On the basis of the results, estimate the probability of getting a total of seven on the two dice.

Set the seed of 4321, as in the first lines of screen (28). Generate 99 integers between 1 and 6 and store them in L1, with randInt(1,6,99) for the TI-83, as in the last lines of screen (28); use seq(iPart 6rand+1,X,1,99,1) for the TI-82.

With ↑ENTRY, change L1 to L2 for the first lines of screen (29). Add these two lists and store the result in L3, with L1+L2→L3, as in the last lines of screen (29).

In the last line of screen (29), notice that the first value of L3 is 4 + 6 = 10; the second value is 3 + 4 = 7. To see the proportion of rows that add to 7, construct a histogram of L3, as in screen (32), with the two set-up screens in (30) and (31).

The last line of screen (32) shows that there are 19 totals of 7. This is 19 of 99, or 19/99 = 0.191919, or about 19% totals of 7 compared to the theoretical 6/36, or 16.7%. We could do this simulation several times and combine the results for an answer that we could have more confidence in.

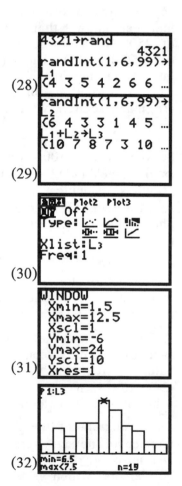

(28)

(29)

(30)

(31)

(32)

COUNTING [pg. 159]

Factorial!

EXAMPLE [pg. 161]: 3! = 3*2*1 = 6

EXAMPLE [pg. 162]: How many different routes are possible if you must visit each of the 50 state capitals?

You can visit any one of the 50 states first. For each state, there are 49 ways of picking the second state, 48 ways of picking the third state, and so on until only one state is left to visit. Thus 50*49*48 . . . 3*2*1 = 50!

Type **50** and then press **MATH<PRB>4:!** Then press **ENTER** for the last line of screen (33) or approximately 3E64, or 3*10^64, or 3 followed by 64 zeros — a lot of possibilities!

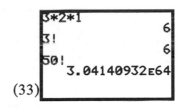

(33)

Permutations [pg. 163]

EXAMPLE [pg 163]: In planning the Monday night prime-time line-up for the NBC television network, an executive must select 6 shows from 30 shows that are available. How many different line-ups are possible?

Type **30**, press **MATH<PRB>2:nPr**, and then type **6**. Press **ENTER** for 427,518,000 possible lineups, as in screen (34).

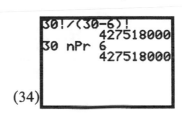

(34)

Combinations [pg 165]

EXAMPLE [pg. 165]: The Board of Trustees at the author's college has nine members. Each year, the board elects a three-person committee to oversee buildings and grounds. Each year, the board also elects a chairperson, vice-chairperson, and secretary. How many different three-person committees are possible?

Type **9**, press **MATH<PRB>3:nCr**, and then type **3**. Press **ENTER** for 84 possible committees, as in screen (35).

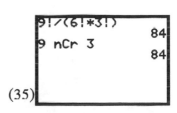

(35)

COMPUTER PROJECT [pg. 176]

Use simulation to estimate the probability of getting at least two people who share the same birthday when randomly selecting 25 people. (See the birthday problem simulation on page 38.)

4

Probability Distributions

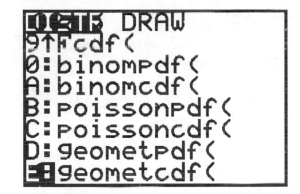

In this chapter you will learn about discrete probability distributions by first plotting the probability histograms and then calculating the mean, variance, and standard deviation of a distribution given in table form. You will also find out how to calculate the binomial and Poisson probabilities on the Home screen and then use the ↑DISTR functions (TI-83 only), which will alleviate the need to calculate probabilities or to look them up in a table.

PROBABILITY DISTRIBUTION BY TABLE [pg. 186]

EXAMPLE [pg. 186]: Suppose that USAir runs 20% of all flights and that all flights have the same chance of crashing. If we let the random variable x represent the number of USAir crashes among seven randomly selected crashes, the probability distribution of the number of crashes can be described by the table at the right.

L1	L2
x	P(x)
0	0.21
1	0.367
2	0.275
3	0.115
4	0.029
5	0.004
6	0+
7	0+

Probability Histogram [pg. 187 and Figure 4-3]

To plot the probability histogram from the table follow these steps:

1. Put the x values in L1 (you could use seq(X,X,0,7,1→ L1) and the P(x) values in L2 (0, not 0+, for the last two values). **Note**: sum L2=1.

2. (TI-83 only) Set up Plot1 and the WINDOW as in screens (1) and (2). Press **TRACE** for the probability histogram in screen (3). From the cell highlighted and centered at 1, we observe that n = 0.367 = P(1).

2A. (For TI-82) Let L3=1000∗L2, similar to what we did for the relative frequency histogram on pages 19 and 20. Adjust the Plot and WINDOW set-ups since the TI-82 needs integers for the frequencies.

Note: From the TI-82 histogram, P(1) = 367/1000 = 0.367.

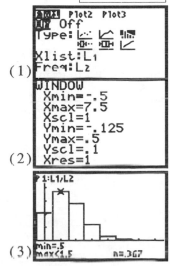

(1)

(2)

(3)

Mean, Variance, and Standard Deviation [pg. 189]

EXAMPLE [pg. 190]: Use the probability distribution model for USAir crashes to find the mean number of USAir crashes (among seven), the variance, and the standard deviation.

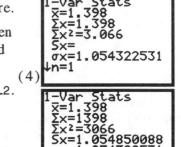

1. Put the x values in L1 and the P(x) values in L2, as before.

2. (TI-83 only) Press **STAT<CALC>1:1–Var Stats L1,L2**, and then **ENTER** for screen (4) with $\bar{x}$ actually being $\mu = 1.398$ and $\sigma = \sigma x = 1.054322531$ so $\sigma^2 = 1.0543^2 = 1.11$.   (4)
   Notice that $n = 1$ equals the sum of the probabilities in L2.

2A. (For TI-82) Let L3=1000✳L2 and then press **STAT<CALC>1:1–Var Stats L1,L3** and **ENTER** for screen (5), with the same results for the mean, standard deviation, and variance as above, but now $n = 1000$.   (5)

BINOMIAL DISTRIBUTION [pg 196]

Binomial Probability Formula [pg. 199]

$$P(x) = nCx * p^x(1 - p)^{n-x} \quad \text{for } x = 0, 1, 2, \ldots, n$$

EXAMPLE [pg. 200]: Given that 10% of us are left-handed, find the probability of having exactly 3 left-handed students in a class of 15; that is, find P(3) given that $n = 15$, $x = 3$, $p = 0.1$, and $q = 0.9$.

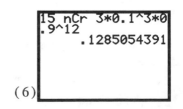

On the Home screen, type and paste

15 nCr **3✳0.1^3✳0.9^12**

and then **ENTER** for 0.1285 = P(3), as in screen (6).

Note: nCr is under **MATH<PRB>**.   (6)

↑DISTR 0:binompdf and ↑DISTR A:binomcdf [pg. 201]

EXAMPLE [pg. 200 (extended to show other possibilities)]:
Given that n = 15, x = 3, p = 0.1, and q = 0.9 find
(a) the probability of <u>exactly 3</u> successes.
(b) the probability of <u>at most 3</u> successes.
(c) the probability of <u>at least 3</u> successes.
(d) the probability of <u>3 to 8</u> successes.

   To get the complete table of values, press
↑DISTR 0:binompdf(**15,0.1** then **ENTER** for screen (7)
with P(0) = .20589, P(1) = .3.... Use the ▶ key to
reveal the other values.

(7)

(**a**) Given that n = 15, x = 3, p = 0.1, and q = 0.9, find
the probability of <u>exactly 3</u> successes.

   Press **↑DISTR 0:**binompdf(**15,0.1,3** and then **ENTER** for
<u>0.1285</u>, as in screen (8). This repeats the solution in
screen (6).

Note: pdf stands for probability density function.

(**b**) Find the probability of <u>at most 3</u> successes, or
P(0) + P(1) + P(2) + P(3).

   Press **↑DISTR A:**binomcdf(**15,0.1,3** and then **ENTER** for
<u>0.944</u>, as in screen (9).

Note: cdf stands for cumulative density function
(cumulative from 0 to x).

Alternatively, you can create a list of P(0) + P(1) + P(2)
+ P(3), as in the first lines of screen (10), and then sum
the values in the list as in the last lines.

Note: sum under **↑LIST** <MATH>.

(**c**) Find the probability of <u>at least 3</u> successes, or
P(3) + P(4) + P(5) + ... + P(14) + P(15)
= 1 − [P(0) + P(1) + P(2)].

   Press **↑DISTR A:**binomcdf(**15,0.1,2** and then **ENTER** for
the first lines of screen (11) for P(0) + P(1) + P(2).
Then type **1 - ↑ANS** and press **ENTER** for <u>0.184061</u>.

(**d**) Find the probability of 3 to 8 successes, or
P(3) + P(4) + P(5) + P(6) + P(7) + P(8) =
[P(0) + P(1) P(2) + P(3) + ... + P(7) + P(8)]
− [P(0) + P(1) + P(2)].

   Use binomcdf(**15,0.1,8**) - binomcdf(**15,0.1,2**) = <u>0.184058</u>
for the solution as in screen (12).

(8)

(9)

(10)

(11)

(12)

Mean, Variance, and Standard Deviation for Binomial Distribution [pg. 210]

EXAMPLE [pg. 210]: USAir crashes can be considered a binomial experiment with n = 7, p = 0.20, and q = 0.80. Given these values compare the results from the simplified formulas ($\mu = n * p$, $\sigma = \sqrt{(n*p*q)}$) with those from using 1-Var Stats.

Now that we know about binomcdf(n,p), we can carry many decimal places than the probability distribution table that started this chapter. Put the x values of 0 to 7 in L1 and the binomial probabilities in L2. (See screen (13).)
Use 1-Var Stats L1,L2 for screen (14), with μ, designated
$\bar{x}$ = 1.4, and σ, or σx = 1.0583.

The above results are in agreement with
$\mu = n * p = 7 * 0.20 = 1.4$ and
$\sigma = \sqrt{(n*p*q)} = \sqrt{(7*0.20*0.80)} = 1.0583$
as given in screen (15).

Note: $\sqrt{}$ at A6.

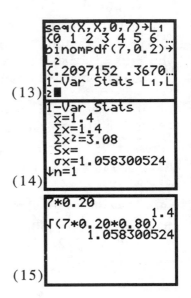

(13)

(14)

(15)

POISSON DISTRIBUTION [pg. 215]

EXAMPLE [pg. 217 (extended to show other possibilities)]: In analyzing hits by V-1 buzz bombs in World War II, South London was subdivided into 576 regions, each with an area of 0.25 sq. km. A total of 535 bombs hit the combined area of 576 regions for an average of μ = 535/576 = 0.929 hits per region. If a region is randomly selected, find (using the Poisson distribution)

(a) the probability that it was hit <u>exactly twice</u>.
(b) the probability that it was hit <u>at most twice</u>.
(c) the probability that it was hit <u>at least twice</u>.
(d) the probability that it was hit <u>from 2 to 6</u> times.

Poisson Probability Formula $P(x) = \mu^x e^{\wedge}(-\mu)/x!$

Given $\mu = 0.929$,

(a) find the probability that it was hit <u>exactly twice</u>.

Type in the first line $2 \rightarrow X : 0.929^X * e^{\wedge} -0.929/X!$
and then press **ENTER** for 0.1704, as in screen (16).

Note: We are connecting two statements with a colon (above the decimal point in the last row). We store 2 in X and calculate $P(2) = \underline{0.1704}$. **e^** comes from pressing ↑**e^X** at A8; **!** comes from **MATH<PRB>**. Engaging ↑**ENTRY** and pressing ▲ ▲ jumps us to the beginning, where we can type another value for X if we like.

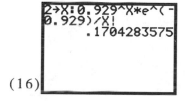

(16)

↑DISTR B:poissonpdf and ↑DISTR C: poissoncdf

Given $\mu = 0.929$,

(a) find the probability that it was hit <u>exactly twice</u>.

Press ↑**DISTR B:poissonpdf(0.929,2)** and then **ENTER** for screen (17), with $P(2) = \underline{0.1704}$ and as above.

(17)

(b) find the probability that it was hit <u>at most twice</u>.

Press ↑**DISTR C:poissoncdf(0.929,2)** and then **ENTER** for screen (18), with $P(0) + P(1) + P(2) = \underline{0.9323}$.

Note: Screen (19) shows another method.

(18)

(c) find the probability that it was hit <u>at least twice</u>,
or $P(2) + P(3) + P(4) + P(5) + ... = 1 - [P(0) + P(1)]$.

Press ↑**DISTR C:poissoncdf(0.929,1)** and then **ENTER** for the first lines of screen (20) for $P(0) + P(1)$.
Then type **1 - ↑ANS** and press **ENTER** for $\underline{0.23814}$,
which is the solution.

(19)

(20)

(d) find the probability that it was hit <u>from 2 to 6</u> times.

$P(2) + P(3) + P(4) + P(5) + P(6) =$
$[P(0) + P(1) + P(2) + ... + P(5) + P(6)] - [P(0) + P(1)]$
or ↑**DISTR C:poissoncdf(0.929,6)** - ↑**DISTR C:poissoncdf(0.929,1)**
$= \underline{0.23809}$ as in screen (21).

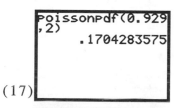

(21)

5

Normal Probability Distribution

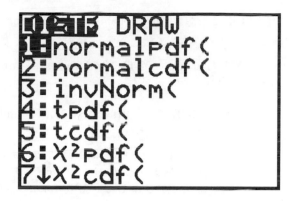

In this chapter you will use the normalcdf and invNorm functions of the TI-83 to calculate areas under a normal curve and to do the inverse, that is, to find a value from a given area or probability. These functions replace the need to use the tables in the text that are necessary for those who have a TI-82. You will again see the importance of the normal distribution when we simulate the central limit theorem and approximate some binomial distributions with a normal distribution.

STANDARD NORMAL DISTRIBUTION [pg. 228]

Finding Probabilities When Given z Scores [pg. 231]

EXAMPLE [pg. 232]: Find the area under the normal curve between z = 0 and z = 1.58.

1. Press ↑**DISTR** 2:normalcdf(**0,1.58** and then **ENTER** for 0.442947, as in screen (1).

   Note: ↑DISTR 2:normalcdf(leftmost or low z, rightmost or upper z).

2. To shade and calculate the area:

   (a) Set up the WINDOW as in screen (2), with all plots off.

   (b) Press ↑**DISTR**<DRAW>1:ShadeNorm(**0,1.58** as in screen (3) and then **ENTER** for screen (4).

   Note: To shade another area, you will need to clear the previous drawing or shading with ↑**DRAW** 1:ClrDraw and then **ENTER** for 'Done.'

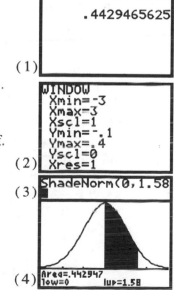

EXAMPLE [pg. 234]: Find the area under the normal curve greater than z = 1.27.

1. Press ↑**DISTR 2**:normalcdf(**1.27,E99** and then **ENTER** for 0.1020, as in screen (5).

   Note: The E for the **E99** is from ↑**EE** (above the comma) and stands for 10^99, or 10 times itself 99 times, for a very large number to the right. Use ⁻**E99** for a value far to the left.

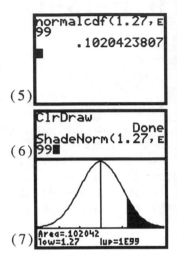

(5)

2. To shade and calculate the area:

   (a) Set up the **WINDOW** as in screen (2), with all plots off.

   (b) Press ↑**DRAW 1**:ClrDraw and then **ENTER** for 'Done' (as in the first two lines of screen (6)) because you need to clear the shading from the previous example.

   (c) Press ↑**DISTR**<DRAW>**1**:ShadeNorm(**1.27,E99** as in screen (6) and then **ENTER** for screen (7).

(6)

(7)

Finding Z Scores When Given Probabilities [pg. 237]

EXAMPLE [pg. 238]: Find the z score that is the 95th percentile, separating the top 5% from the bottom 95% of the area under the standard normal distribution.

1. Press ↑**DISTR 3**:invNorm(**0.95** and then **ENTER** for 1.64485, as in the first two lines of screen (8).

   Note: ↑**DISTR 3**:invNorm(area to left of z).

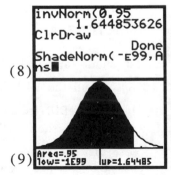

2. To shade the desired area:

   (a) Set up the **WINDOW** as in screen (2), with all plots off.

   (b) Press ↑**DRAW 1**:ClrDraw and then **ENTER** for 'Done' (as in the middle lines of screen (8)) because you need to clear the shading from the previous examples.

   (c) Press ↑**DISTR**<DRAW>**1**:ShadeNorm(⁻**E99,**↑**ANS** for the last two lines of screen (8) and then press **ENTER** for screen (9).

   Note: **ANS** is in the last row of the keyboard. You could also have typed in the z value of 1.64485 directly.

(8)

(9)

NONSTANDARD NORMAL DISTRIBUTION
Finding Probabilities [pg. 242]

EXAMPLE [pg. 243]: If American women's heights are normally distributed with a mean = μ = 63.6 in. and a standard deviation = σ = 2.5 in., find the percentage of American women with heights between 64.5 in. and 72 in.

A. Press ↑**DISTR** 2:normalcdf(**64.5,72,63.6,2.5** and then **ENTER** for 0.359034, as in screen (10).

 Note: ↑DISTR 2:normalcdf(low value,upper value,μ,σ).

B. To shade and calculate the area:

 1. Set up the WINDOW as in screen (11) as follows with all plots off:

 Xmin = -3*σ + μ = -3*2.5 + 63.6 = 56.1

 Xmax = 3*σ + μ = 3*2.5 + 63.6 = 71.1

 Xscl = 1

 Ymin = -0.1/σ = -0.1/2.5 = -0.4

 Ymax = 0.4/σ = 0.4/2.5 = 0.16

 Yscl = 0

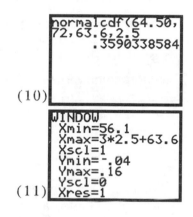

(10)

(11)

 Note: Screen (11) shows that the above calculations can be done in the WINDOW editor, but the calculation for Xmax shown will be calculated to 71.1 when you press **ENTER** or use the Down arrow.

 2. Press ↑**DISTR<DRAW>**1:ShadeNorm(**64.5,72,63.6,2.5** as in screen (12) and then **ENTER** for screen (13).

 Note: To shade another area using the same window, you will need to clear the previous drawing or shading with ↑**DRAW** 1:ClrDraw and then **ENTER** for 'Done.'

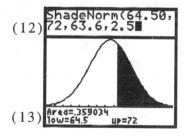

(12)

(13)

Finding Scores [pg. 249]

EXAMPLE [pg. 250]: Find P$_{90}$ (the 90th percentile) of women's heights if the heights are normally distributed with μ = 63.6 and σ = 2.5.

Press ↑**DISTR** 3:invNorm(**0.90,63.6,2.5** and then **ENTER** for 66.8 in. as in screen (14). Only 10% of women are taller than 5 ft. 6.8 in.

Note: ↑DISTR 3:invNorm(area to left of score,μ,σ)

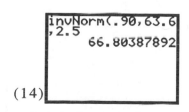

(14)

CENTRAL LIMIT THEOREM [pg. 255]

As the sample size increases, the sampling distribution of sample means approaches a normal distribution.

Here you will see the plausibility of the central limit theorem through a simulation that finds the means of the last four digits of social security numbers.

1. On the Home screen, set your seed so that you can duplicate your results, with **4321 STO►** rand, as on page 8 (step 4) and the first two lines of screen (15).

2. Set up the following sequences, similar to what was done in Chapter 3 and in screen (15). (Do not forget the last entry feature of **↑ENTRY**, as shown on pages 9 and 10.)

   seq(randInt(0,9),X,1,50,1 **STO► L1 ENTER**
   seq(randInt(0,9),X,1,50,1 **STO► L2 ENTER**
   seq(randInt(0,9),X,1,50,1 **STO► L3 ENTER**
   seq(randInt(0,9),X,1,50,1 **STO► L4 ENTER**

   Note: On the TI-82, replace the randInt(0,9) with iPart 10rand.

(15)
```
4321→rand
              4321
seq(randInt(0,9)
,X,1,50,1)→L1
{5 3 7 5 1 9 9 …
seq(randInt(0,9)
,X,1,50,1)→L2∎
```

3. Find the mean of each row of these lists using the following (on the Home screen):
   (L1+L2+L3+L4)/4 **STO► L5 ENTER**

4. Press **STAT 1:**Edit for screen (16). Notice that the first four values in the first row are relatively large and their mean = (5 + 9 + 6 + 6)/4 = 6.5, whereas the first four values in the fifth row are relatively small with a mean of 2.25. It would be very unlikely that all four values are 9 or all four values are 0. It is much more likely that there are large and small values that will average out to a middle value.

(16)
L1	L2	L3	3	L4	L5	3
5	9	6		6	6.5	
3	6	3		0	3	
7	9	4		2	5.5	
5	1	0		4	2.5	
1	0	5		3	2.25	
9	0	7		9	6.25	
9	6	1		2	4.5	

L3(1)=6

5. To summarize the data, set up the WINDOW and Plot1 as in screens (17) and (18). Press **TRACE** and ► for the histogram of screen (19). The distribution of the 50 values in L1 is fairly uniform, with about 5 of each of the 10 digits 0, 1, 2, 3, 4, 5, 6, 7, 8, and 9.

(17)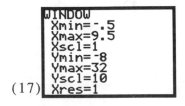

(18)

(19)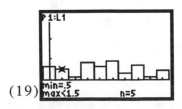

Note:The 200 values in L1, L2, L3, and, L4 can be regenerated and plotted as in screens (20) and (21) with "Xlist" of Plot1 changed to L6. Again you will see a fairly uniform distribution, but this time with about 20 values in each cell. The 200 values could have been combined in L6, instead of regenerated, (as explained on page 15) with augment(L1, L2 **STO►** L6 **ENTER**, augment(L6, L3 **STO►** L6 **ENTER**, augment(L6, L4 **STO►** L6 **ENTER**.

(20)

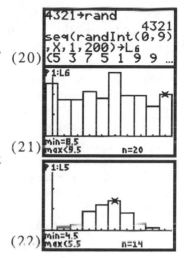

6. Set up Plot1 for a histogram of the means stored in L5. **TRACE** reveals the histogram of screen (22). Notice that this histogram is much more normally shaped, with no means of 0 (the first cell) or 9 (the last cell), as we had conjectured. Most of the means are lumped near the center at 4, 5, and 6 and predicted by the central limit theorem.

(21)

(22)

EXAMPLE [pg. 258]: Given that the population of men has normally distributed weights, with a mean of 173 lbs. and a standard deviation of 30 lbs., find the probability that

(a) if 1 man is randomly selected, his weight is greater than 180 lbs.
(Answer: 0.408 as in first three lines of screen (23).)

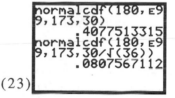

(b) if 36 different men are randomly selected, their mean weight is greater than 180 lbs.
(Answer: 0.0808 as in the last three lines of screen (23) using a standard error of $30/\sqrt{36}$.)

(23)

It is much less likely that the mean of 36 values is 7 lbs. above the population mean than one randomly selected value.

NORMAL DISTRIBUTION AS AN APPROXIMATION TO THE BINOMIAL DISTRIBUTION [pg. 266]

If $np \geq 5$ and $n(1 - p) \geq 5$, then the binomial random variable is approximately normally distributed with the mean and standard deviation given as $\mu = np$ and $\sigma = \sqrt{(np(1-p))}$

EXAMPLE [pg. 267]: Assume that your college has an equal number of qualified male and female applicants, and assume that 64 of the last 100 newly hired employees are men. Use the normal distribution to estimate the probability of getting <u>at least 64</u> men if each hiring is done independently with no gender discrimination.

With n * p = 100 * 0.5 = 50 > 5 and n * q = 100*(1 − 0.50) = 50 > 5, we could use the normal distribution to approximate the binomial with μ = np = 50 and σ = √ (100*0.50*0.50) = 5. Using the continuity correction to get the area for P(64) + P(65) + P(66) + ... + P(100), we need to start with 63.50 and continue out to the right for a normal approximation of 0.0035 as in the first lines of screen (24).

You can check this answer since P(64) + P(65) + P(66) + ... + P(99) + P(100) = 1 − [P(0) + P(1) + P(2) + ... + P(62) + P(63)] and the exact binomial probability can be calculated as we did in Chapter 4 and in the last lines of screen (24) for 0.0033.

Note: All of the problems in section 5.6 of the text can be checked using the binomcdf function except for 18, which has n = 1000; binomcdf works only for n up to 999.

(24)

EXERCISE 18 [pg. 274] : Estimate the probability that of any 1000 randomly selected videotaped shows, at least 700 are from major networks if 66% of shows taped are from the major networks.

Since n * p = 1000 * 0.66 = 660 > 5 and n * q = 1000 * 0.34 = 340 > 5, we can use a normal approximation to the binomial with μ = n * p = 660 and σ = √ (1000 * 0.66 * 0.34) = 14.758.

(25)

To approximate P(700) + P(701) + ... + P(999) + P(1000) and include 700, we start at 699.5 for an approximate answer of 0.0037 as in screen (25).

(26)

If we try to solve this with binomcdf as in screen (26) we get a DOMAIN error as in screen (27).

Note: Exercise 9 on page 278 has N = 2500 and will require a normal approximation.

(27)

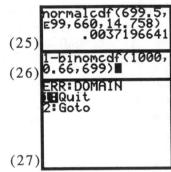

COMPUTER PROJECT [pg. 280]

In screen (10) of this chapter we found that 35.9% of women have heights between 64.5 in. and 72 in. if the heights of women are normally distributed with a mean of 63.6 in. and a standard deviation of 2.5 in. Randomly generate 100 heights using the randNorm function under **MATH**<PRB> to see if we are close to these results.

First set the seed as in the first two lines of screen (28) so that you can repeat these results. Randomly generate 100 values from a simulated normal distribution, and then put these values in order in list L1 as in screen (28).

Using **STAT 1**:Edit with ▼, go to the 70th row for the first value above 64.5 (64.5171793 as in the bottom line of screen (29)). Continuing down the list, we find the last, or 100th, value is 71.818274 as in the last line of screen (30), which is less than 72 in. Since from 70 to 100 is 31 values, we have 31% of the values between 64.5 and 72 compared to the theoretical value of 35.9%.

Try different seeds or larger sample sizes to vary these results.

(28)

```
123→rand
                    123
randNorm(63.6,2.
5,100)→L1
(62.26033074 65…
SortA(L1
               Done
```

(29)

L1	L2	L3	1
64.311			
64.331			
64.382			
64.384			
64.419			
64.476			
64.517			

L1(70) =64.5171793…

(30)

L1	L2	L3	1
67.099			
67.151			
67.456			
68.136			
70.891			
71.818			

L1(100) =71.818274…

6

Estimates and Sample Sizes

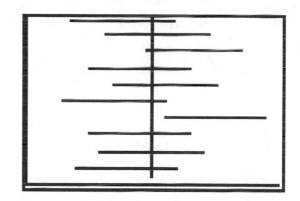

In this chapter you will learn how to estimate population parameters by taking one simple random sample from the population. The **STAT<TESTS>** menu of the TI-83 will be used, but the calculations will also be shown on the Home screen. These calculations will clarify what equation is being used; they are also suitable for the TI-82. To use the **STAT<TESTS>** functions, you can input the data in two ways:

(1) Using summary statistics such as $\bar{x}$ and n (see screen (1))
(2) Using the raw data stored in a list (see screen (7)).

ESTIMATING A POPULATION MEAN: LARGE SAMPLES [pg. 288]

EXAMPLE [pg. 294]: For the body temperature data [Table 6-1], we have n = 106, $\bar{x}$ = 98.20, and s = 0.62. For a 0.95 degree of confidence, use these statistics to find the margin of error, E, and the confidence interval for μ.

1. Press **STAT<TESTS>** 7:ZInterval for a screen similar to screen (1). Make sure that the input line (Inpt:) has Stats highlighted by using the ▶ key and pressing **ENTER**. Since n = 106 >30, use ▼ and let σ = s = 0.62. Also, input $\bar{x}$, n, and the confidence level (C-Level) as in screen (1). (1)

```
ZInterval
 Inpt:Data Stats
 σ:.62
 x̄:98.2
 n:106
 C-Level:.95
 Calculate
```

2. Highlight Calculate in the last line of screen (1) and press **ENTER** for screen (2) with the 95% confidence interval of (98.082,98.318), or 98.082 < μ < 98.318. You can calculate the margin of error by taking the larger value of the confidence interval and subtracting $\bar{x}$, or E = 98.318 − 98.2 = 0.118. So you can also express the 95% confidence interval as 98.2 ± 0.118. (2)

```
ZInterval
 (98.082,98.318)
 x̄=98.2
 n=106
```

3. You can do the Home screen calculations using
E = $z_{\alpha/2}*\sigma \div \sqrt{n}$ = 1.96*0.62/$\sqrt{}$ (106) **STO▶** E
as shown in screen (3).

Note: You can find $z_{\alpha/2}$ = 1.96 in Table A-2 of the text, but (3)

```
1.96*0.62/√(106)
→E
          .118030658
98.2-E
          98.08196934
98.2+E
          98.31803066
```

you can also calculate it with the TI-83, as is shown in Chapter 7, screen (4).

ESTIMATING A POPULATION MEAN: SMALL SAMPLES AND THE STUDENT t DISTRIBUTION [pg. 305]

Using Summary Statistics

EXAMPLE [pg. 309]: Analysis of 12 damaged Dodge Viper cars from simulated typical collisions results in repair costs having a distribution that appears to be bell-shaped, with a mean of $\bar{x} = \$26,227$ and a standard deviation of $s = \$15,873$. Find the 95% interval estimate of μ, the mean repair cost for all Dodge Vipers involved in collisions.

1. Press **STAT**<TESTS> **8:TInterval** for a screen similar to screen (4). Make sure that the input line (Inpt:) has Stats highlighted by using the ▶ key and pressing **ENTER**. Input $\bar{x}$, Sx, n, and the C-Level as in screen (4).

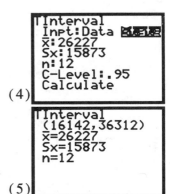

(4)

2. Highlight Calculate in the last line of screen (4), and press **ENTER** for screen (5) with the 95% confidence interval of ($16,142, $36,312), or $16,142 < \mu < \$36,312$. You can calculate the margin of error by taking the larger value of the confidence interval and subtracting $\bar{x}$, or E = 36312 − 26227 = 10085. So you can also express the 95% confidence interval as $26,227 ± $10,085.

(5)

3. You can do the Home screen calculations using

$$E = t_{\alpha/2}*s \div \sqrt{n} = 2.201*15873/\sqrt{(12)} \text{ STO▸ E}$$

as shown in screen (6).
Note: You can find $t_{\alpha/2} = 2.201$ in Table A-3 of the text, but you can also calculate it with the TI-83, as is shown in Chapter 7, screens (11), (12), and (13) for a similar case.

(6)

Using Raw Data List

EXERCISE 10 [pg. 313]: Assume the 10 body temperatures below (stored in L1) are randomly selected from a distribution that is normally distributed. Construct the 95% confidence interval for the mean of all body temperatures.

L1 = {98.6 98.6 98.0 98.0 99.0 98.4 98.4 98.4 98.4 98.6}

1. Press **STAT**<TESTS> **8:TInterval** for a screen similar to screen (7). Make sure that the input line (Inpt:) has Data highlighted. Input list L1 with Freq set at 1 and C-Level at 0.95.

2. Highlight Calculate and press **ENTER** for screen (8).

(7)

The 95% confidence interval is (98.229, 98.651), or 98.229 < μ < 98.651. E = 98.651 − 98.44 = 0.21, so you could also write the 95% confidence interval as 98.44   ± 0.21

Note: The mean and standard deviation of the data are calculated by the function and given in screen (8).

(8)

```
TInterval
 (98.229,98.651)
x̄=98.44
Sx=.2951459149
n=10
```

ESTIMATING A POPULATION PROPORTION [pg. 315]

EXAMPLE [pg. 317]: In a survey of 1068 Americans, 673 stated that they had telephone answering machines. Find a 95% confidence interval estimate of the population proportion of all Americans who have answering machines.

1. Press STAT<TESTS> A:1-PropZInt and then set up a screen similar to the one in screen (9). Input x (the number with answering machines), n (the number in the sample), and the confidence level (C-Level) as in screen (9).

(9)

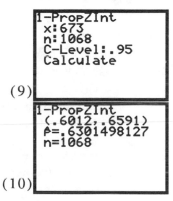

```
1-PropZInt
x:673
n:1068
C-Level:.95
Calculate
```

2. Highlight Calculate in the last line of screen (9), and press ENTER for screen (10), with a point estimate of p̂ = 0.63 = 63% and a 95% confidence interval of (0.6012, 0.6591), or 60.1% < p < 65.9%. You can calculate the margin of error by taking the larger value of the confidence interval and subtracting p̂, or E = .6591 − .6301 = 0.029. So you can also express the 95% confidence interval as 0.630 ± 0.029.

(10)

```
1-PropZInt
 (.6012,.6591)
p̂=.6301498127
n=1068
```

3. The home screen calculations are shown in screens (11) and (12) using the following formulas:

$\hat{p}$ = 673/1068 = 0.630 = P

$E = z_{\alpha/2} * \sqrt{(P*(1-P)/n)}$

  = 1.96*$\sqrt{(0.63*(1-0.63)/1068)}$ = 0.029

(11)

(12)

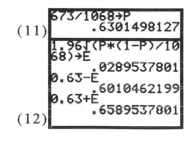

```
673/1068→P
        .6301498127
1.96√(P*(1-P)/10
68)→E
        .0289537801
0.63-E
        .6010462199
0.63+E
        .6589537801
```

Note: If $\hat{p}$ were given as 0.63 and n = 1068, you could not input x:0.63*1068 for 672.84 in screen (9). Doing so would cause a domain error since x must be an integer. You must first round to the nearest integer, or 673.

ESTIMATING A POPULATION VARIANCE [pg. 325]

EXAMPLE [pg. 330]: The Hudson Valley Bakery makes doughnuts. Twelve doughnuts are randomly selected from the production line and weighed, with the results given below (in ounces and stored in L1). Construct a 95% confidence interval for σ to check whether the standard deviation meets quality control standards by being less than 0.06 oz.

{3.43   3.37   3.58   3.50   3.68   3.61   3.42   3.52   3.66   3.50   3.36   3.42}

1. With the data in L1, press **STAT<CALC>1:1-Var Stats L1** for screen (13), with Sx = s = 0.109 for the 12 values above.

```
1-Var Stats
 x̄=3.504166667
 Σx=42.05
 Σx²=147.4811
 Sx=.1090836487
 σx=.1044396423
↓n=12
```
(13)

2. From Table A-4 of the text, for $\alpha = 0.05$ with $12 - 1$ degrees of freedom you will find that $\chi_L^2 = 3.816$ and $\chi_R^2 = 21.920$.

   Note: You will see how the TI-83 can find these critical values in the last section of Chapter 7.

3. $(n - 1)s^2/\chi_R^2 < \sigma^2 < (n - 1)s^2/\chi_L^2$ [Formula on pg. 329]
   $(12 - 1)0.109^2/21.920 < \sigma^2 < (12 - 1)0.109^2/3.816$
   $$0.0059621807 < \sigma^2 < 0.0342481656$$
   $$0.07721 < \sigma < 0.18506.$$

```
(12-1)*0.109²/21
.920
        .0059621807
√(Ans)
        .0772151582
```
(14)

The standard deviation appears to be too large, and the quality control manager must take corrective action to make the doughnut weights more consistent.

COMPUTER PROJECT (THE BOOTSTRAP METHOD FOR CONFIDENCE INTERVALS) [pg. 339]

L3 = {2.9   564.2   1.4   4.7   67.6   4.8   51.3   3.6   18.0   3.6}

Using the sample data given above, and storing it in L3, construct a 95% confidence interval estimate of the population mean μ by using the following method.

(a) Create 500 new samples, each of size 10, by selecting 10 scores with replacement from the 10 sample scores above.

(b) Take the mean of each sample.

(c) Put the sample means in order.

You can carry out steps (a), (b), and (c) on the TI-83 with the program listed in screens (15) and (16).

(15)
(16)

Note: The TI-82 is limited to 99 samples, and randInt(1,10 must be replaced with iPart 10rand +1.

Screens (17) and (18) show the output of running the program after setting a seed.

(d) Calculate $P_{2.5}$ and $P_{97.5}$ as in screen (19) for the estimate $P_{2.5} < \mu < P_{97.5} = 6.475 < \mu < 184.73$.

To calculate the 95% confidence interval estimate for the population standard deviation σ, change **mean** in the program to **stdDev**, which is found under ↑LIST<MATH>7.

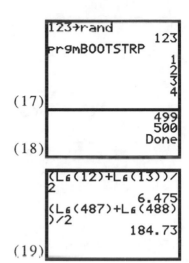

(17)

(18)

(19)

7

Hypothesis Testing

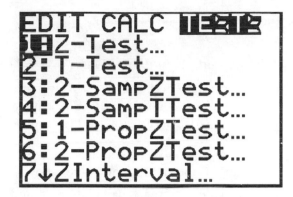

In this chapter you'll use the **STAT<TESTS>** functions of the TI-83 to do one-sample statistical tests. Calculations that are suitable for the TI-82 will also be shown, on the Home screen, and will help to clarify the procedure used. You should refer to the main text, however, to find out the requirements necessary to make the test valid and the procedure to follow to include all the steps your instructor has asked for.

To use the **STAT<TESTS>** functions, you can input the data in two ways:

(1) Using summary statistics such as $\bar{x}$ and n (see screen (1))

(2) Using the raw data stored in a list (see screen (9))

The output of these tests gives the p-value but also the test statistic, so you can use the traditional method of hypothesis testing, checking the test statistic against the critical value. For the TI-83, you don't need to use the tables in the text to find the critical values, as you'll soon see. You can use the Draw option for the output (see screen (3)), which can help clarify a test.

TESTING A CLAIM ABOUT A MEAN: LARGE SAMPLES [pg. 360]

Traditional Method (Two-Tail Test)

EXAMPLE 1a [pg. 361]: Using the sample data of temperatures given at the beginning of the chapter (n = 106, $\bar{x}$ = 98.20, s = 0.62) and a 0.05 significance level, test the claim that the mean body temperature of healthy adults is equal to

98.6°F $H_0: \mu = 98.6$ $H_1: \mu \neq 98.6$

1. Press **STAT<TESTS>1:Z-Test** for a screen similar to screen (1). Make sure that the input line (Inpt:) has Stats highlighted by using ▶ and pressing **ENTER**. Use ▼ to input μ_0:98.6, and since n = 106 > 30, use ▼ and let $\sigma = s = 0.62$. Also input $\bar{x}$ and **n** as in screen (1), highlight the the alternate hypothesis $\mu:\neq \mu_0$, and then press **ENTER**.

(1)

2. Highlight 'Calculate' and press **ENTER** for screen (2).

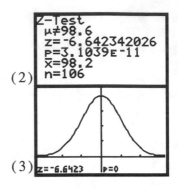

(2)

If you had highlighted Draw in the last line of screen (1) and then pressed **ENTER**, you would have obtained screen (3). Both screens (2) and (3) reveal that the test statistic z = ⁻6.6423. In screen (3) this value is far to the left of the given standard normal distribution since -3.5 is about the minimum value shown on the z axis.

(3)

3. The Home screen calculation could be done using

$$z = (\bar{x} - \mu_0)/(\sigma/\sqrt{(n)}) = (98.2 - 98.6)/(0.62/\sqrt{(106)})$$
$$= ⁻6.6423$$

as in the top lines of screen (4).

4. The critical z value can be found in Table A-2 of the text or calculated as in the last few lines of screen (4) using invNorm (with ↑**DISTR 3** as in Chapter 5) for $z_{\alpha/2} = \pm 1.96$.

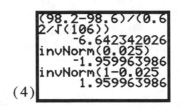

(4)

Since ⁻6.6423 < ⁻1.96 we reject the null hypothesis and conclude that μ is statistically significantly less than 98.6.

P-value Method [pg. 366]

EXAMPLE 1b [pg. 368]: Use the p-value method to test the claim of the preceding example.

Following steps 1 and 2 above, you obtained screen (2), which gives a p-value = 3.1039E⁻11 = 0.000000000031039, given in screen (3) as p=0 to four decimal places.

Since the p-value < 0.05, we reject the null hypothesis, as before.

Note: When the p-value is part of the output, there is no need to calculate or look up the critical values. This makes it a more efficient approach than the traditional method.

Note: The area shaded in each tail of the distribution in screen (3) is half of 3.1039E⁻11, but this is too small to show on the screen. Screen (16) gives another example of a two-tail test, in which the area is split between both tails.

Traditional Method (One-Tail Test)

EXAMPLE 2a [pg. 364]: In its advertisements, the Jack Wilson Health Club claims that "you will lose weight after only two days of the Jack Wilson diet and exercise program." To test the claim, 33 people who had followed the program were checked for weight loss. They had lost, on average, 0.37 lb. with a standard

deviation of 0.98 lb. Use a 0.05 level of significance to test
whether the advertised claim is true.

$$H_0: \mu \leq 0 \qquad H_1: \mu > 0$$

1. Press **STAT**<**TESTS**>1:**Z-Test** for a screen similar to
   screen (5). Make sure that the input line (Inpt:) has Stats
   highlighted by using ▶ and pressing **ENTER**. Use ▼ to
   input μ_0:0, and since n = 33 > 30, use ▼ and let $\sigma = s = .98$.
   Also input $\bar{x}$ and n as in screen (5), highlight the
   Alternate Hypothesis $\mu :> \mu_0$, and press **ENTER**.

2. Highlight **Calculate** and press **ENTER** for screen (6), or
   Draw and **ENTER** for screen (7). Both screens give the
   test statistic of z = 2.1689, with the area to the right of
   this value shaded in the standard normal curve of
   screen (7).

3. Home screen calculations would be similar to those
   in step 3 of Example 1a.

4. The critical z value is 1.645, as shown in screen (8)
   using ↑**DISTR 3** as in Chapter 5. Since 2.1689 > 1.645, we
   reject H_0 and conclude that there was a statistically
   significant mean weight loss.

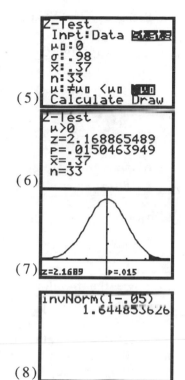

(5)

(6)

(7)

(8)

P-value Method (One-Tail Test)

EXAMPLE 2b [pg. 370]: Use the p-value method to test
the claim of the preceding example.

Following steps 1 and 2 above, you obtained screens (6)
and (7), which give a p-value = 0.015 < 0.05, so we reject the
null hypothesis as before. Notice that 0.015 of the area under
the standard normal distribution is shaded in the upper
tail in screen (7).

TESTING A CLAIM ABOUT A MEAN: SMALL SAMPLES [pg. 379]

EXAMPLE [pg. 382]: The seven scores listed below are axial loads (in pounds) for
the first sample of seven 12-oz. aluminum cans. (Store these scores in L1.) At the
0.01 level of significance, test the claim that this sample comes from a population
with a mean that is greater than 165 lbs.

L1 = { 270   273   258   204   254   228   282}

$H_0: \mu \le 165$ $H_1: \mu > 165$

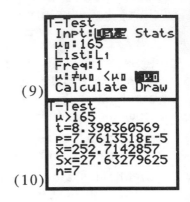

1. Press **STAT**<**TESTS**>**2:T-Test** for a screen similar to screen (9). Make sure that the input line (Inpt:) has Data highlighted by using ◄ and pressing **ENTER**. Use ▼ to input μ_0:165, paste L1 for the data List, and set Freq to 1. Highlight the alternate hypothesis $\mu:> \mu_0$, and press **ENTER**.

(9)

2. Highlight Calculate and press **ENTER** for screen (10), with the test statistic of t = 8.3984 and the p-value = 0.0000776. The sample statistics $\bar{x}$ and Sx = s are also calculated.

(10)

3. Home screen calculations are as follows:

   $t = (\bar{x} - \mu_0)/(s/\sqrt{(n)}) = (252.714 - 165)/(27.633/\sqrt{(7)})$
   $= 8.398$

4. Since the p-value = 0.0000776 <0.01, we reject H_0 and conclude that μ is significantly greater than 165 lbs.

The critical t value with 6 degrees of freedom and $\alpha = 0.01$ (3.143) can be found in Table A-3 of the text or calculated on the TI-83 as follows:

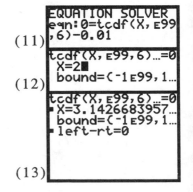

1. Press **MATH** 0:Solver for a screen similar to screen (11). (If the screen looks more like screen (12), press ▲.) To get the screen to look like screen (11), paste tcdf(from ↑**DISTR 5**. You will solve for an X that has 0.01 of area to its right under a t-distribution with df = 6.

(11)

(12)

2. Press **ENTER** for screen (12). Type **2** as first guess for X=2 since you want to be in the right tail.

3. With the cursor flashing on the X=2 line, press **ALPHA SOLVE** (above **ENTER**), and wait for the calculation (be patient) and X = 3.1427, as in screen (13).

(13)

Since the critical t = 3.143 < 8.3984 (the test statistic of screen (10)) we reject H_0 and conclude (as before) that μ is significantly greater than 165 lbs.

TESTING A CLAIM ABOUT A PROPORTION [pg. 389]

EXAMPLE [pg. 392]: In a consumer taste test, 100 regular Pepsi drinkers are given blind samples of Coke and Pepsi: 48 of these subjects preferred Coke. At the 0.05 level of significance, test the claim that Coke is preferred by 50% of Pepsi drinkers who participate in such blind taste tests.

$$H_0: p = 0.5 \qquad H_1: p \neq 0.5$$

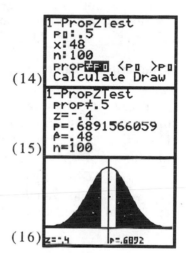

1. Press **STAT<TESTS>5:1-PropZTest** for a screen similar to screen (14). Complete the information as in screen (14).

2. Highlight Calculate and press **ENTER** for screen (15), or Draw and **ENTER** for screen (16), with the test statistic z = ⁻0.4 and a p-value = 0.6892. (Screen (16) has 0.6892/2 = 0.3445 shaded in each tail because this is a two-tail test.) Screen (15) also gives the calculation of $\hat{p} = x / n = 0.48$.

3. Home screen calculations are as follows:

$$z = (\hat{p} - p_0)/(\sqrt{(p_0 * (1 - p_0)/n)}$$
$$= (0.48 - 0.50)/\sqrt{(0.5*0.5/100)} = {}^-0.40$$

4. Since the p-value = 0.6892 > 0.05, we fail to reject H_0 and conclude that there is no significant difference in the proportion who prefer one cola over the other. (Likewise, z = 0.40 < the critical value 1.96 leads to the same conclusion.)

TESTING A CLAIM ABOUT A STANDARD DEVIATION OR A VARIANCE [pg. 398]

EXAMPLE [pg. 400]: Altimeters are manufactured with errors that are normally distributed with a mean of 0 ft and a standard deviation of 43.7 ft. After the installation of new production equipment, 30 altimeters were randomly selected from the new line. This sample group had errors with a standard deviation of s = 54.7 ft. Use a 0.05 significance level to test the claim that the new altimeters have a standard deviation different from the old value of 43.7 ft.

H_0: $\sigma = 43.7$ H_1: $\sigma \neq 43.7$

1. Calculate the test statistic from the following equation given in the text [pg. 399].

   $$\chi 2 = (n - 1)\ s^2/\sigma^2 = (30 - 1)54.7^2/43.7^2 = 45.437$$

2. Calculate the p-value as in screen (17) with $\chi2$cdf from ↑**DISTR 7** for p = 0.05334. This is a two-tail test.

3. Since the p-value = 0.0533 > 0.05, we fail to reject the null hypothesis. There is not sufficient evidence to support the claim that the standard deviation is different from 43.7 ft.

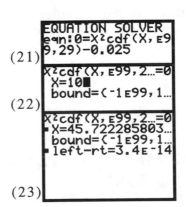

(17)

The critical values from Table A-4 of the text are $\chi^2_L = 16.047$ and $\chi^2_R = 45.722$ with degrees of freedom = 29. (You can calculate these critical values with the **MATH** Solver, similar to what was done for the critical t-value in screens (11), (12), and (13). Screens (18), (19), and (20) are for calculating χ^2_L, and screens (21), (22), and (23) are for calculating χ^2_R.) Since the test statistic 45.437 is between the two critical values (and thus not in the critical region), we fail to reject the null hypothesis as before.

(18)
```
EQUATION SOLVER
eqn:0=X²cdf(0,X,
29)-0.025
```

(19)
```
X²cdf(0,X,29)…=0
 X=1█
 bound=(-1E99,1…
```

(20)
```
X²cdf(0,X,29)…=0
▪X=16.047071701…
 bound=(-1E99,1…
▪left-rt=-1E-14
```

(21)
```
EQUATION SOLVER
eqn:0=X²cdf(X,E9
9,29)-0.025
```

(22)
```
X²cdf(X,E99,2…=0
 X=10█
 bound=(-1E99,1…
```

(23)
```
X²cdf(X,E99,2…=0
▪X=45.722285803…
 bound=(-1E99,1…
▪left-rt=3.4E-14
```

8

Inferences from Two Samples

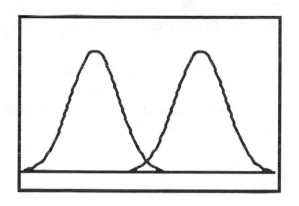

In Chapter 6 and 7 you found confidence intervals and tested hypotheses for data sets that involved only one sample from one population. In this chapter you will learn how to calculate test statistics and confidence intervals for data sets that involve two samples from two populations. As in Chapters 6 and 7 you will use **STAT<TESTS>** functions. To complete the problems in this Chapter successfully, you need to be familiar with Data list and sample Statistics input, and Calculate and Draw output possibilities, from the examples in Chapters 6 and 7.

INFERENCES ABOUT TWO MEANS: DEPENDENT SAMPLES [pg. 418]

EXAMPLE [pg. 420]: Subjects are tested for reaction times with their left and right hands. (Only right-handed subjects were used.) The results (in thousandths of a second) are given below. Use a 0.05 level of significance to test the claim that there is a difference between the mean of the right-hand and left-hand reaction times.

Subject	A	B	C	D	E	F	G	H	I	J	K	L	M	N
Right hand L1	191	97	116	165	116	129	171	155	112	102	188	158	121	133
Left hand L2	224	171	191	207	196	165	177	165	140	188	155	219	177	174
Difference L3=L1-L2	-33	-74	-75	-42	-80	-36	-6	-10	-28	-86	33	-61	-56	-41

$$H_0: \mu_d = 0 \qquad\qquad H_1: \mu_d \neq 0$$

Put the right-hand times in L1 and the left-hand times in L2. Store the difference of L1 - L2 in L3 (i.e., highlight L3 in the **STAT 1:**Editor, type L1 − L2, and then press **ENTER**) to obtain the results in the last line of the table above.

1. Press **STAT<TESTS>2:T-Test** and set up the resulting screen as in screen (1).

2. Highlight **Calculate** and press **ENTER** for screen (2), with t = $^-$4.794 and the p-value = 0.00035.
   Note that $\bar{d} = \bar{x} = ^-42.5$ and s_d = Sx = 33.17, which could also be calculated with **STAT[CALC]1:1-Var Stats L3**.

3. The Home screen calculation is as follows:
   $$t = (\bar{d} - \mu)/(s/\sqrt{(n)}) = (^-42.5 - 0)/(33.173/\sqrt{(14)}) = ^-4.794$$

4. Since p-value = 0.00035 is less than the significance level 0.05, we would reject the null hypothesis. The right hand is significantly faster than the left hand.

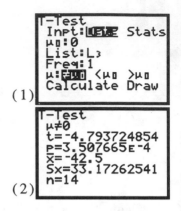

(1)

(2)

Confidence Intervals for μ_d [pg. 423]

EXAMPLE [pg. 422]: Use the sample data from the preceding example to construct a 95% confidence interval estimate of μ_d.

1. Press **STAT<TESTS>8:TInterval** and set up the resulting screen as in screen (3).

2. Highlight **Calculate** and press **ENTER** for screen (4), with the 95% confidence interval equal to ($^-61.65$, $^-23.35$), or $^-61.65 < \mu_d < ^-23.35$.

   The margin of error E can be calculated by taking the difference between $\bar{d}$ and the larger interval value.
   E = $^-23.35 - (^-42.5) = 19.5$, so we can also express the 95% confidence interval as $^-23.35 \pm 19.15$.

3. The Home screen calculation uses $t_{\alpha/2} = 2.160$ (with 13 degrees of freedom), so
   E = $t_{\alpha/2}*s/\sqrt{(n)} = 2.160*33.173/\sqrt{(14)} = 19.15$ as above.

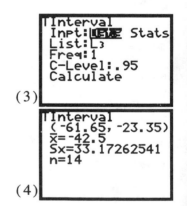

(3)

(4)

INFERENCES ABOUT TWO MEANS: INDEPENDENT AND LARGE SAMPLES [pg. 429]

EXAMPLE [pg. 431]: Use a 0.01 significance level to test the claim that cans 0.0109 in. thick have a lower mean axial load (in pounds) than cans that are 0.0111 in. thick. The original data are listed in Data Set 15 [Appendix B], and the summary statistics are listed below.

0.0109 in. Cans	0.0111 in. Cans
$n_1 = 175$	$n_2 = 175$
$\bar{x}_1 = 267.1$	$\bar{x}_2 = 281.8$
$s_1 = 22.1$	$s_2 = 27.8$

H_0: $\mu_1 \geq \mu_2$ (or $\mu_1 - \mu_2 \geq 0$) H_1: $\mu_1 < \mu_2$

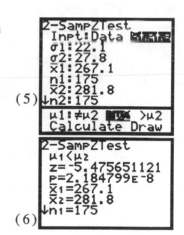

1. Press **STAT<TESTS>3:2-SampZTest** and set up the resulting screen as in screen (5), using $\sigma_1 = s_1$ and $\sigma_2 = s_2$ because of the large sample sizes

2. Highlight Calculate and press **ENTER** for screen (6), with z = -5.476 and the p-value = 0.0000000218.

3. The Home screen calculation is as follows.

   $z = ((\bar{x}_1 - \bar{x}_2) - (\mu_1 - \mu_2))/\sqrt{(\sigma_1{}^2/n_1 + \sigma_2{}^2/n_2)}$

   $= (267.1 - 281.8)/\sqrt{(22.1^2/175 + 27.8^2/175)}$

   $= -5.476$

   (5)

   (6)

4. Since the p-value = 0.0000000218 is less than the significance level 0.05, we would reject the null hypothesis. The thinner can withstood a (statistically) significantly smaller mean axial load than did the thicker can.

Confidence Interval [pg. 432]

EXAMPLE [pg. 432]: Using the sample data given in the preceding example, construct a 99% confidence interval estimate of the difference between the means of the axial loads of the 0.0109 in. cans and the 0.0111 in. cans.

1. Press **STAT<TESTS>9:2-SampZInt** and set up the resulting screen as in screen (7) using $\sigma_1 = s_1$ and $\sigma_2 = s_2$ because of the large sample sizes.

2. Highlight Calculate and press **ENTER** for screen (8), with the <u>99%</u> confidence interval equal to (-21.62, -7.785), or $-21.62 < \mu_1 - \mu_2 < -7.785$.

You can use $\bar{x}1 - \bar{x}2 = 267.1 - 281.8 = -14.7$ to find the margin of error E by subtracting it from the larger interval value. $E = -7.785-(-14.7) = 6.915$, so you can also write the 99% confidence interval as -14.7 ± 6.915.

(7)

3. The Home screen calculation uses $z_{\alpha/2} = 2.575$ for

$E = z_{\alpha/2}* \sqrt{(\sigma1^2/n1 + \sigma2^2/n2)}$

$= 2.575* \sqrt{(22.1^2/175 + 27.8^2/175)} = 6.9$ as above.

(8)

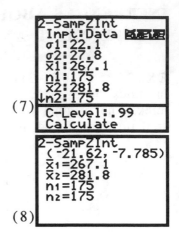

COMPARING TWO VARIANCES [pg. 439]

EXAMPLE [pg. 442]: For the aluminum can data, use a 0.05 significance level to test the claim that the samples come from populations with the same variance.

0.0109 in. Cans	0.0111 in. Cans
n1 = 175	n2 = 175
$\bar{x}1 = 267.1$	$\bar{x}2 = 281.8$
s1 = 22.1	s2 = 27.8

$H_0: \sigma_1 = \sigma_2$ (or $\sigma_1/\sigma_2 = 1$) $H_1: \sigma_1 \neq \sigma_1$

1. Press **STAT<TESTS>D:2-SampFTest** and set up the resulting screen as in screen (9).

2. Highlight Calculate and press **ENTER** for screen (10), with F = 0.631967 and the p-value = 0.002618235.

Note: If you prefer your F value to be greater than 1, reverse the order in which you enter the data, as in screen (11), for the results in screen (12), with F = 1.582359 but the same p-value = 0.002618235. (Also note that 1/0.631967 = 1.582359.)

(9)

(10)

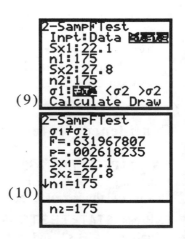

3. The Home screen calculation is as follows:

$F = s1^2/s2^2 = 22.1^2/27.8^2 = 0.631967$ with
$(n1 - 1, n2 - 1) = (174, 174)$ degrees of freedom or

$F = s2^2/s1^2 = 27.8^2/22.1^2 = 1.582349$ with

$(n2 - 1, n1 - 1) = (174, 174)$ degrees of freedom.

4. Since the p-value = 0.002618 is less than the significance level 0.05, we would reject the null hypothesis. The thicker cans have (statistically) significantly larger variation in axial loads than do the thinner cans.

(11)

(12)

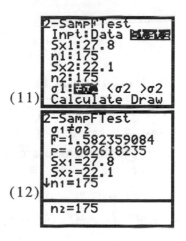

INFERENCES ABOUT TWO MEANS: INDEPENDENT AND SMALL SAMPLES [pg. 447]

Assumptions
1. The two samples are independent.
2. The two samples are randomly selected from normally distributed populations.
3. At least one of the two samples is small ($n \leq 30$).

Case 1: Both Population Variances Are Known [pg. 448]

Because of the assumptions above, this will be a z-test, even with the small sample size, and will be treated exactly as the large-sample case described on page 71.

Case 2: Equal Variances (We Fail to Reject $\sigma_1^2 = \sigma_2^2$) [pg. 450]

If the variances are equal, you can pool (combine) the sample variances to get a weighted average of s_1^2 and s_2^2, which is the best possible estimate of the variance σ^2 that is common to both populations.

EXAMPLE: [pg. 450]: Use the sample statistics that follow at the 0.05 level of significance to test the claim that the mean amount of nicotine in filtered king-size cigarettes is equal to the mean amount of nicotine for nonfiltered king-size cigarettes. (All measurements are in milligrams.)

Filtered kings: $n_1 = 21$, $\bar{x}_1 = 0.94$, $s_1 = 0.31$
Nonfiltered kings: $n_2 = 8$, $\bar{x}_2 = 1.65$, $s_2 = 0.16$

Preliminary F Test (H_0: $\sigma_1 = \sigma_2$)

1. Press **STAT<TESTS>D:2-SampFTest** and set up the resulting screen as in screen (13).

2. Highlight Calculate and press **ENTER** for screen (14), with the p-value = 0.0799.

3. Since $0.0799 > 0.05$, we fail to reject H_0: $\sigma_1 = \sigma_2$. This is Case 2, where we <u>pool</u> the variance.

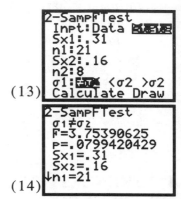

(13)

(14)

Test H_0: $\mu_1 = \mu_2$ (or $\mu_1 - \mu_2 = 0$) H_1: $\mu_1 \neq \mu_2$

1. Press **STAT<TESTS>4:2-SampTTest** and set up the resulting screen as in screen (15). Notice that in the second-to-last line, Yes is highlighted for Pooled.

2. Highlight Calculate and press **ENTER** for screen (16), with t = ⁻6.126, p-value = 0.0000015198, and sp = Sxp = 0.27897 (for $sp^2 = 0.0778$).

(15)

3. The Home screen calculation is as follows:
$$sp^2 = ((n_1-1)s_1{}^2 + (n_2-1)s_2{}^2)/((n_1-1) + (n_2-1))$$

$$= (20*0.31^2 + 7*0.16^2)/(20 + 7) = 0.0778$$

$$t = ((\bar{x}_1 - \bar{x}_2) - (\mu_1 - \mu_2))/\sqrt{(sp^2/n_1 + sp^2/n_2)}$$

$$= (0.94 - 1.65)/\sqrt{(0.27897^2/21 + 0.27897^2/8)}$$

$$= {}^-6.127$$

(16)

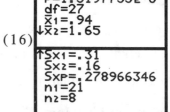

4. Since the p-value = 0.0000015198 is less than the significance level 0.05, we would reject the null hypothesis. The filtered cigarettes have significantly less nicotine than the nonfiltered ones.

Confidence Interval

EXAMPLE [pg. 451]: Using the data in the preceding example, construct a 95% confidence interval estimate of $\mu_1 - \mu_2$.

1. Press **STAT**<TESTS>**0:2-SampTInt** and set up the resulting screen as in screen (17).

2. Highlight Calculate and press **ENTER** for screen (18), with the 95% confidence interval equal to $(-0.9478, -0.4722)$ or $-0.9478 < \mu_1 - \mu_2 < -0.4722$.

   $\bar{x}1 - \bar{x}2 = 0.94 - 1.65 = -0.71$ to find the margin of error E by subtracting it from the larger interval value. $E = -0.4722 - (-0.71) = 0.2378$, so you can also write the 95% confidence interval as -0.71 ± 0.2378.

3. The Home screen calculation uses $t_{\alpha/2} = 2.052$ (with $n1 + n2 - 2 = 21 + 8 - 2 = 27$ degrees of freedom for

   $E = t_{\alpha/2}* \sqrt{(sp^2/n1 + sp^2/n2)}$

   $= 2.052* \sqrt{(0.27897^2/21 + 0.27897^2/8)} = 0.2378$ as above.

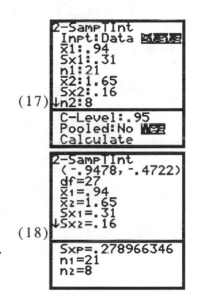

(17)

(18)

Case 3: Unequal Variances (Reject $\sigma_1^2 = \sigma_2^2$) [pg. 452]

EXAMPLE [pg. 453]: Use the following sample data at the 0.05 level of significance to test the claim that the mean amount of <u>tar</u> in filtered king-size cigarettes is less than the mean amount of tar for nonfiltered king-size cigarettes. (All measurements are in milligrams.)

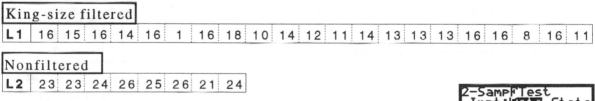

King-size filtered																				
L1	16	15	16	14	16	1	16	18	10	14	12	11	14	13	13	13	16	16	8	16	11

Nonfiltered							
L2	23	23	24	26	25	26	21	24

Preliminary F Test (H_0: $\sigma_1 = \sigma_2$)

1. Press **STAT**<TESTS>**D:2-SampFTest** and set up the resulting screen as in screen (19).

2. Highlight Calculate and press **ENTER** for screen (20), with the p-value = 0.0385.

3. Since $0.0385 < 0.05$, we reject H_0: $\sigma_1 = \sigma_2$. This is Case 2, where we do <u>not pool</u> the variance.

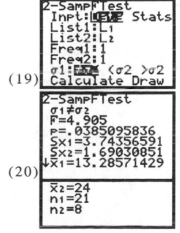

(19)

(20)

Test H$_0$: $\mu_1 \geq \mu_2$ (or $\mu_1 - \mu_2 \geq 0$)   H$_1$: $\mu_1 < \mu_2$

1. Press **STAT<TESTS>4:2-SampTTest** and set up the resulting screen as in screen (21). Notice that in the second-to-last line No is highlighted for Pooled.

2. Highlight Calculate and press **ENTER** for screen (22), with t = $^-$10.585 and the p-value = 0.00000000003312.

3. The Home screen calculation is as follows:

   $t = ((\bar{x}1 - \bar{x}2) - (\mu_1 - \mu_2))/\sqrt{(s1^2/n1 + s2^2/n2)}$

   $= (13.285 - 24)/\sqrt{(3.74^2/21 + 1.69^2/8)} = {}^-10.59$

4. Since the p-value = 0.000$^+$ is less than the significance level 0.05, we would reject the null hypothesis. The filtered cigarettes have significantly less tar than the nonfiltered ones.

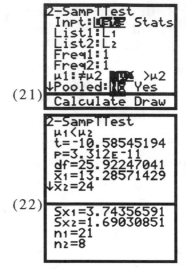

(21)

(22)

Confidence Interval

EXAMPLE [pg. 454]: Use the data in the preceding example to construct a 95% confidence interval estimate of $\mu_1 - \mu_2$.

1. Press **STAT<TESTS>0:2-SampTInt** and set up the resulting screen as in screen (23).

2. Highlight Calculate and press **ENTER** for screen (24), with the 95% confidence interval equal to ($^-$12.8, $^-$8.633), or $^-$12.8 < $\mu_1 - \mu_2$ < $^-$8.633.

   You can use $\bar{x}1 - \bar{x}2$ = 13.2857 − 24 = $^-$10.714 to find the margin of error E by subtracting it from the larger interval value. E = $^-$8.633 − ($^-$10.714) = 2.081, so you can also write the 95% confidence interval as $^-$10.714 ± 2.081.

3. The Home screen calculation uses

   $E = t_{\alpha/2} * \sqrt{(s1^2/n1 + s2^2/n2)}$

   $= 2.060 * \sqrt{(3.74^2/21 + 1.69^2/8)} = 2.08$ as above.

$t_{\alpha/2} = 2.060$ is from df = 25 (conservative estimate from 25.922 of screen (24)). You can use Formula 8-1 of the text [pg. 453] to calculate degrees of freedom (df):

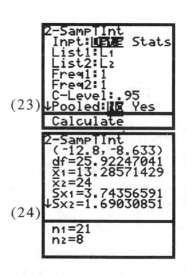

(23)

(24)

$$df = (A+B)^2/[A^2/(n_1 - 1) + B^2/(n_2 - 1)]$$

where $A = s_1^2/n_1$ and $B = s_2^2/n_2$.

See the calculation in screen (25).

Note: The text used $df = 8 - 1 = 7$ with a critical t of 2.391 for the conservative (wider) confidence interval of $(^-13.1 , ^-8.3)$

(25)

```
3.74²/21→A
        .6660761905
1.69²/8→B
        .3570125
(A+B)²/(A²/20+B²
/7)
        25.91435201
```

INFERENCES ABOUT TWO PROPORTIONS [pg. 462]

EXAMPLE [pg. 463]: Johns Hopkins researchers conducted a study of pregnant IBM employees. Among 30 who worked with glycol ethers, 10 (or 33.3%) had miscarriages; among 750 who were not exposed to glycol ethers, 120 (or 16%) had miscarriages. At the 0.01 significance level, test the claim that the miscarriage rate was greater for women exposed to glycol ethers.

Exposed to glycol ethers: $n1 = 30$ $x1 = 10$ $\hat{p}_1 = 10/30 = 0.333$

Not exposed to glycol ethers: $n2 = 750$ $x2 = 120$ $\hat{p}_2 = 120/750 = 0.160$

$H_0: p_1 \le p_2$ (or $p_1 - p_2 \le 0$) $H_1: p_1 > p_2$

1. Press **STAT<TESTS>6:2-PropZTest** and set up the resulting screen as in screen (26).

2. Highlight **Calculate** and press **ENTER** for screen (27), with $z = 2.498$ and p-value = 0.006245.

3. The Home screen calculation is as follows:

   $p = (x1 + x2)/(n1 + n2) = (10 + 120)/(30 + 750) = 0.1667$

   $z = ((\hat{p}_1 - \hat{p}_2) - (p_1 - p_2))/\sqrt{(p*q/n1 + p*q/n2)}$
   $= (0.333 - 0.160)/\sqrt{(0.1667*0.8333/21 + 0.1667*0.8333/8)}$
   $= 2.493$

4. Since the p-value = 0.0062 is less than the significance level 0.01, we would reject the null hypothesis. The miscarriage rate is significantly greater for women exposed to glycol ethers.

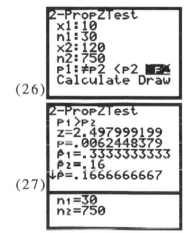

(26)

(27)

Confidence Interval   [pg. 466]

EXAMPLE [pg. 467]: Use the data in the preceding example to construct a 99% confidence interval for the difference between the two population proportions.

1. Press **STAT<TESTS>B:2-PropZInt** and set up the resulting screen as in screen (28).

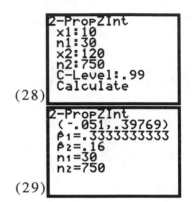

(28)

2. Highlight **Calculate** and press **ENTER** for screen (29), with the 99% confidence interval equal to (-0.051, 0.398), or $-0.051 < p_1 - p_2 < 0.398$.

   You can use $\hat{p}_1 - \hat{p}_2 = 0.333 - 0.16 = 0.173$ to find the margin of error E by subtracting it from the larger interval value. $E = 0.398 - 0.173 = 0.225$, so you can also write the 99% confidence interval as 0.173 ± 0.225.

(29)

3. The Home screen calculation uses
   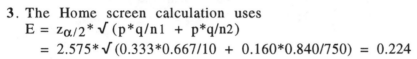
   $E = z_{\alpha/2} * \sqrt{(p*q/n1 + p*q/n2)}$

   $= 2.575 * \sqrt{(0.333*0.667/10 + 0.160*0.840/750)} = 0.224$

9

Correlation and Regression

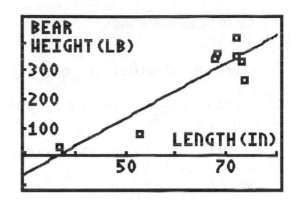

Our study of correlation and regression will be aided by the **STAT<TESTS>E**: LinRegTTest function. This function is used for linear regression with Y as a straight-line function of X. Program **A2MULREG** applies to multiple regression where Y is a function of X1, X2, . . ., Xn and supplements the computer output given in the text.

Chapter Problem [pg. 481]: Researchers have studied bears by anesthetizing them and obtaining their vital measurements. The table below gives the lengths and weights of a sample of eight bears. Put the lengths in L1 and the weights in L2. On the basis of this data, find out whether there appears to be a relationship between the length of a bear and its weight. If there is, what is the relationship? If a researcher anesthetizes a bear and uses a tape measure to find that the bear is 71.0 in. long, how do you use that length to predict the bear's weight?

x: Length (in.) L1	53	67.5	72	72	73.5	68.5	73	37
y: Weight (lb.) L2	80	344	416	348	262	360	332	34

SIMPLE LINEAR REGRESSION AND CORRELATION

Scatter Plot [pg. 483]

1. Enter data into L1 and L2.

2. Set up Plot1 for a scatter plot as in screen (1).

3. Press **ZOOM 9**:ZoomStat and **TRACE** for the scatter plot in screen (2).

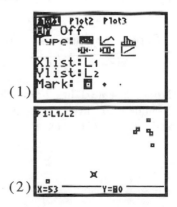

(1)

(2)

"The points in the figure seem to follow a pattern, so we might conclude that there is a relationship between the length of a bear and its weight. This conclusion is largely subjective because it is based on our perception of whether a pattern is present." [pg. 483]

Correlation   Coefficient   r

EXAMPLE [pg. 486]: Using the previous data, find the value of the linear correlation coefficient r.

1. Press **STAT<TESTS>E**:LinRegTTest and set up the resulting screen as in screen (3). (Y1 is pasted from **VARS<YVARS>1**: Function 1: Y1.)

(3)

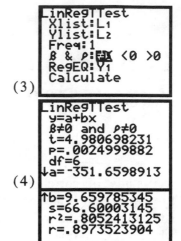

2. Highlight **Calculate** in the last line, and press **ENTER** for the two screens of output (4) with r = 0.89735 in the last line.

(4)

Note: There are other ways of finding. r on the TI-83, but this way will be the most productive.

Note: For the TI-82, engage **STAT<CALC>9**:LinReg(a+bx) **L1, L2 ENTER** for a screen with the linear correlation coefficient r = 0.89735.

Formal   Hypothesis   Test   of   the   Significance   of   r   [pg. 490]

EXAMPLE [pg. 492]: Using the data above, test the claim that there is a linear correlation between lengths and weights of bears.

$$H_0: \rho = 0 \quad H_1: \rho \neq 0$$

Using the first screen of the output of **STAT<TESTS>E**: LinRegTTest (screen (4)), we see the test statistic t = 4.981 and the p-value = 0.0025 < 0.05, indicating significant positive correlation.

Screen (5) uses the following formula for t from the text $t = r / \sqrt{((1 - r^2)/(n - 2))}$ [pg. 490].

(5)

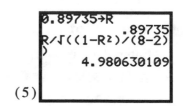

Note: The text has t = 4.971 [pg. 492] but uses r = 0.897 (rounded to three decimal places); screen (5) used r = 0.89735, and screen (4) uses r = 0.8973523904.

Regression   [pg. 500]

EXAMPLE [pg. 503]: For the chapter problem, you used the given data (x = lengths and y = weights of bears) to find that the linear correlation coefficient of r = 0.897 indicates that there is significance linear correlation. Now find the regression equation of the straight line that relates x and y.

Repeat the procedure from earlier in the chapter.

1. Press **STAT<TESTS>E:LinRegTTest** and set up the resulting screen as in screen (6).

2. Highlight **Calculate** in the last line, and press **ENTER** for the two screens of output (7) with a = b_0 = -351.66 and

   b = b_1 = 9.66, thus $\hat{y}$ = -351.66 + 9.66x.

3. Set up Plot1 as in screen (1). The regression equation (**RegEQ**) is automatically stored and turned on as Y1 in the **Y =** editor by step 1. All other plots must be turned off. Press **ZOOM 9:ZoomStat** and then **TRACE** for the least square regression line plotted with the scatter plot of points as in screen (8).

Note: On the TI-82 (1) press **STAT<CALC> 9:LinReg(a+bx) L1,L2** and then **ENTER** for a screen with a = b_0 and b = b_1. (2) Under the **Y=** editor and next to Y1=, press **VARS 5:Statistics <EQ> 7:RegEq** to paste the regression equation for plotting. (3) Press **ZOOM 9:ZoomStat**.

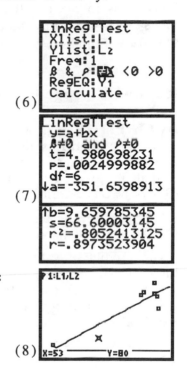

(6)

(7)

(8)

Predictions   [pg. 505]

EXAMPLE [pg. 506]: Using the sample data of the chapter problem, you found that there is significant linear correlation between lengths and weights of bears, and you also found the regression equation as above. If a bear is measured and is found to be 71.0 in. long, predict its weight.

1. While in **TRACE** mode as in screen (8), press ▲ for the cursor to move from a data point to the regression line, as in screen (9). (The equation listing starts at the top of the screen.)

2. Type **71**, and a large X = 71 shows in the bottom line as in screen (10). Press **ENTER** for screen (11), with a predicted y value of 334.2 lbs. as in the bottom line of

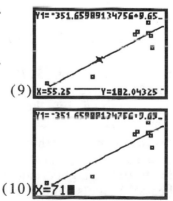

(9)

(10)

screen (11).

You can also calculate the predictive value on the Home screen (and on the TI-82) by typing in the regression equation as in the top lines of screen (12) or by pasting Y₁ to the Home screen and then typing **(71)** and pressing **ENTER**, as in the last line of screen (12). (Y₁ comes from **VARS<YVARS>1**: Function 1: Y₁ on the TI-83 and from **↑YVARS 1**: Function 1: Y₁ on the TI-82.)

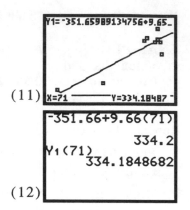

(11)

(12)

Residuals   [pg. 509]

DEFINITION [pg. 509]: For a sample of paired (x, y) data, a **residual** is the difference $y - \hat{y}$ between an observed sample y value and the value of y that is predicted from the regression equation. (**STAT<TESTS>E:LinRegTTest** has been used.)

EXAMPLE: Use the first point of the bear data, x = 53, y = 80, to find its residual.

The predictive value can be found as in the first two lines in screen (13). Subtracting this number from the given y (or 80) as in the third and fourth lines (using Ans in the last row of the keyboard) gives the residual of ⁻80.3087.

(13)

The **STAT<TESTS>E:LinRegTTest** function automatically creates the list of residual called RESID. Paste this to the Home screen from ↑**LIST<NAMES>** (as explained on page 12 (step 6)), and then press **ENTER** for the results in the last lines of screen (13). You can see the other residuals by using the ▶ key.

Coefficient   of   Determination r^2 [pg. 516]

EXAMPLE [pg. 517]: Referring to the bear measurements, find the percentage of the variation in y (weight) that can be explained by the regression line.

The coefficient of determination is $r^2 = 0.897^2 = 0.805$ (as calculated in the top lines of screen (14)) with r from screens (4), (7), or (15). That is, 80.5% of the total variation in bears' weights can be explained by the variation in the bears' heights; the other 19.5% is attributable to other factors.

(14)

Note: If **STAT**<**TESTS**>**E:LinRegTTest** has been used, the last lines of screen (14) show a variation on

r^2 = explained variation/total variation

   = (total variation − unexplained variation)/total variation

   = 1 − unexplained variation/total variation

   = 1 − sum(LRESID^2)/(Sy^2*(n−1))

with sum from ↑**LIST**<**MATH**>, LRESID from ↑**LIST**<**NAMES**>, the **x^2** key is at A6 for squaring, and Sy from **VARS** 5:Statistics<**XY**>6:Sy.

(15)

```
LinRegTTest
y=a+bx
β≠0 and ρ≠0
t=4.980698231
P=.0024999882
df=6
↓a=-351.6598913
```
```
↑b=9.659785345
s=66.60003145
r²=.8052413125
r=.8973523904
```

Note: The output of **STAT**<**TESTS**>**E:LinRegTTest** is in screen (15) to show that r^2 in the second-to-the-last line is part of that output.

Standard Error of Estimate [pg. 518]

EXAMPLE [pg. 518]: Find the **standard error of estimate** for the bear measurement data of the preceding examples. (**STAT**<**TESTS**>**E:LinRegTTest** has been used.)

From the third-last line of screen (15), you see that $se = s = 66.60$ (or $s = \sqrt{(sum(LRESID^2)/(n-2))}$ as in screen (16) with sum from ↑**LIST**<**MATH**> and LRESID from ↑**LIST**<**NAMES**>).

(16)

```
√(sum(LRESID²)/(
8-2))
        66.60003145
```

Note: On the TI-82 with x in L1 and y in L2, store Y1(L1) in L3 (with the RegEQ in Y1) and the residuals, or L2 − L3, in L4, and then $se = \sqrt{(sum(L4^2)/(8-2))}$.

Prediction Intervals for y [pg. 519]

EXAMPLE [pg. 519]: The chapter problem in previous sections, you have seen that there is significant linear correlation and that when x = 71.0, the predicted y value is 334.2 lbs. Construct a 95% prediction interval for the weight of a bear that is 71.0 in. long. This will give you a sense of the reliability of the estimate of 334 lbs. From Table A-3 of the text, $t\alpha/2 = 2.447$ using 8 − 2 degrees of freedom. (**STAT**<**TESTS**>**E**: LinRegTTest has already been used.)

1. The first lines of screen (17) show the length (71 in.) stored in X, 2.447 stored in T, and the predictive value (334.2 in this example) stored in Y, with Y1 (X) **STO**▶Y. These setup values are tied together with colons (above

(17)

```
71→X:2.447→T:Y1(
X)→Y
        334.1848682
T*s√(1+1/n+n(X-x̄
)²/(n*Σx²-(Σx)²)
)→E
        175.5351864
```

the decimal key). Pressing **ENTER** reveals Y = 334.185 from the regression equation Y₁.

2. Lines 4, 5, and 6 of screen (17) show the calculation of E using:

s from **VARS 5**:Statistics<TEST>0:s
n and x̄ from **VARS 5**:Statistics<XY>
Σx² and Σx from **VARS 5**:Statistics<Σ>.

Pressing **ENTER** reveals E = 175.54.

3. Screen (18) reveals the 95% prediction interval of 158.65 < y < 509.72 lbs.
That range is relatively large. One factor contributing to the large range is the small sample size of 8.

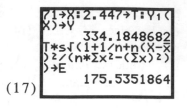

(17)

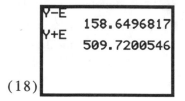

(18)

Note: The predictive interval for different lengths (X) and levels (T) are easily calculated using the last-entry feature. First, recall the top input lines of screen (17) so you can make changes. Then press **ENTER** and recall the lines that calculate E in screen (17).

Note: Calculating predictive intervals for simple linear regression is automated in the program of the next section.

Note: On the TI-82, you would also need to store the standard error of estimate as S and use S instead of s in the equation because s is not calculated.

MULTIPLE REGRESSION [pg. 523]

So far, the examples have involved the relationship between the lengths and weights of bears. Let's extend this by predicting a bear's weight using other information, such as head length, chest size, and neck size. The following table shows data from anesthetized male bears. [pg. 525]

Y or C1 WEIGHT	C2 AGE	C3 HEADL	C4 HEADW	C5 NECK	C6 LENGTH	C7 CHEST
80	19	11	5.5	16	53	26
344	55	16.5	9	28	67.5	45
416	81	15.5	8	31	72	54
348	115	17	10	31.5	72	49
262	56	15	7.5	26.5	73.5	41
360	51	13.5	8	27	68.5	49
332	68	16	9	29	73	44
34	8	9	4.5	13	37	19

Program A2MULREG is introduced below, and its output is compared to the Minitab output. But first we must store the data into the matrix [D]. (The availability of program A2MULREG is given in the Appendix. Be sure your copy is protected as explained with the warning on page 115.)   One important point to remember about data input is that the dependent variable Y, WEIGHT in this example, must be in the first column of [D]. Minitab has no such restriction.

Enter Data into Matrix [D]

The data can be entered from a computer or from another TI-83 (see the Appendix). The following method is for entering the data from the keyboard.

Press **MATRX** [EDIT] **4**:[D] and then **8 ENTER 7 ENTER** for screen (19). You might have to use the **DEL** key when inputting the size of the matrix (8 rows by 7 columns), depending on the size of the previous matrix. Although you may have different values in your matrix, leave them as is because you are going to enter values over them.

(19)

With the cursor in the first cell of the first row (as shown in screen 19), enter the data row by row by typing **80 ENTER 19 ENTER 11 ENTER 5.5 ENTER 16 ENTER 53 ENTER 26 ENTER** for the first row of the data table. With the last **ENTER**, the cursor will return to the first slot in the second row. Enter the other seven rows in the same manner to the last value, as shown in screen (20).

(20)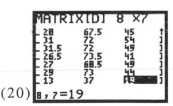

Minitab Output [pg. 525]: Verify the Minitab output with program A2MULREG (for a model with the two independent variables HEADL in C3 and LENGTH in C6) for the following important components:

1. The multiple regression equation
   WEIGHT = -374 + 18.8 HEADL + 5.87 LENGTH

2. Adjusted $R^2 = 0.759$

3. The overall significance of the multiple regression equation or p = 0.012

Program A2MULREG

1. Press **PRGM**<NAMES> and use ▼ to highlight A2MULREG. Press **ENTER**, and prgmA2MULREG is pasted to the Home screen. Press **ENTER** again for screen (22), which

(21)

(22)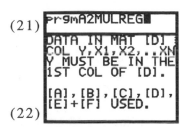

reminds you how to enter data in matrix [D] and alerts you to the fact that data in other matrices would be lost.

Note: Press **ON** to quit if your data is not in martix [D].

2. Press **ENTER** for screen (23) which gives you three options. Press **1** for MULT REGRESSION.

3. The next screen (24) requires the number of independent variables (2) and the column numbers in [D] in which they are located (3 and 6).

4. Pressing **ENTER** after the 6 above gives screen (25), with the overall significance of the multiple regression equation: **P = 0.012**, $R^2 = 0.828$, and R^2 (adjusted), or **R-SQ(ADJ) = 0.7592**.

5. Press **ENTER** for screen (26). Use the CONST and COEFFs to come up with the regression equation:

   WEIGHT = -374.3 + 18.820 HEADL + 5.875 LENGTH

6. Press **ENTER** for screen (27), which gives the option to QUIT; then do quit.

Note: The MAIN MENU of screen (27) gives the options of doing confidence and predictive intervals and of doing residual analysis.

Correlation Matrix

EXAMPLE [pg. 529]: What is the best regression equation to use (the greatest R^2) if only a single independent variable is to be used?

1. When you run program A2MULREG, one of the first screens (screens (23) and (28)) gives the option of calculating the correlation matrix.

Note: Pressing **ENTER** from the Home screen will restart the program for screen (22) if the last thing done on the Home screen was to quit this program.

2. Press **2**:CORR MATRIX, but be aware of the busy calculating indicator in the upper-right corner of the screen—these calculations take a while. The partial output is given in screen (29). (You can view the rest of the output by using the cursor control keys.) The first column is duplicated in the table with headings at the right.

(23)

```
MULT REG+CORR
1:MULT REGRESSIO
2:CORR MATRIX
3:QUIT
```

(24)

```
HOW MANY IND VAR
?2
COL. OF VAR.
                 1
?3
COL. OF VAR.
                 2
?6█
```

(25)

```
    DF  SS
RG  2   113142.268
ER  5   23505.7315
    F=12.03
    P=.012
  R-SQ=.828
  (ADJ).7592
S=68.56490584
```

(26)

```
B0=-374.3034756
CL COEFF / T   P
3  18.82040176
      .81    .453
6  5.874757415
      1.16   .299
```

(27)

```
MAIN MENU
1:CONF+PRI INTER
2:RESIDUALS
3:NEW MODEL
4:QUIT
```

(28)

```
MULT REG+CORR
1:MULT REGRESSIO
2:CORR MATRIX
3:QUIT
```

(29)

```
[[1     .814 .88...
[.814 1      .87...
[.884 .879 1   ...
[.885 .896 .96...
[.971 .906 .95...
[.897 .802 .91...
[.992 .839 .87...
```

	WEIGHT
WEIHGT	1
AGE	0.814
HEADL	0.884
HEADW	0.885
NECK	0.971
LENGTH	0.897
CHEST	0.992

CHEST is the variable that has the greatest linear correlation with WEIGHT, with r = 0.992, or $r^2 = 0.992^2 = 0.984$. NECK is the variable with the next-largest linear correlation coefficient.

3. Run program A2MULREG as in screens (22) and (23) for screen (30). Since we selected **1** independent variable (in column **7** for CHEST), we are given a plot option in screen (31) for the scatter plot in screen (32). Press **ENTER** for screen (33).

Note: Pressing **2** in screen (31) results in screen (33) without a screen (32).

4. Press **ENTER** again for screen (34) which reveals the regression equation

WEIGHT = -194.61 + 11.42*CHEST

Note: Pressing **ENTER** after screens (34) or (26) gives the MAIN MENU of screens (35) and (27). Pressing the **2:RESIDUALS** option of screen (35) gives the options on screen (36), whose plot options lead to screen (37). Quitting the program with option **5** in screen (36) or option **6** in screen (37) gives a list and locations of the residual diagnostics that are calculated (see screen (38)).

(30)
```
HOW MANY IND VAR
?1
COL. OF VAR.      1
?7
```

(31)
```
SCATTER PLOT
1:YES PLOT
2:NO
```

(32)
```
                    P3
         ▫  ▫
      ▫
   P
      R
X=26 ───── Y=80
```

(33)
```
       DF SS
RG 1    134338.134
ER 6    2309.86589
       F=348.95
       P=0.000
       R-SQ=.9831
       (ADJ).9803
S=19.62084729
```

(34)
```
B0=-194.6104038
CL COEFF / T    P
7 11.41554505
      18.68  0.000
```

(35)
```
MAIN MENU
1:CONF+PRI INTER
2:RESIDUALS
3:NEW MODEL
4:QUIT
```

(36)
```
RES OF STAND RES
1:RESIDUAL PLOT
2:STAND.RES.PLOT
3:DURBIN WATSON
4:MAIN MENU
5:QUIT
```

(37)
```
PLOT OF RESID
1:VS YHAT.
2:VS AN IND VAR.
3:VS ROW NUMBER.
4:PREVIOUS MENU
5:MAIN MENU
6:QUIT
```

(38)
```
UNDER STAT EDIT
Y IN LYVAL,THEN
LYHAT, LRES, LSRES
LLEVER, LCOOKD
FOR LEVERAGE AND
COOK DISTANCE
             Done
```

10

Multinomial Experiments and Contingency Tables

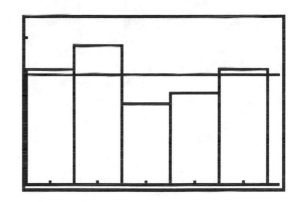

In this chapter you will use the STAT Editor or spreadsheet to calculate the χ^2 statistic for multinomial experiments and the **STAT<TESTS>C:χ^2-Test** function to do contingency table analysis.

MULTINOMIAL EXPERIMENTS: GOODNESS-OF-FIT [pg. 542]

EXAMPLE [pg. 548]: Mars, Inc., claims that its M&M® candies are distributed with the color percentages of 30% brown, 20% yellow, 20% red, 10% orange, 10% green, and 10% blue. The colors of the M&M®s listed in Data Set 11 of Appendix B are summarized in the table below. Using the sample data and a 0.05 significance level, test Mars's claim about the color distribution of M&M® candies.

	Brown	Yellow	Red	Orange	Green	Blue
Observed frequency (L1)	33	26	21	8	7	5
Expected proportion (L3)	0.3	0.2	0.2	0.1	0.1	0.1

To Calculate Chi-square Statistic $\Sigma[(O-E)^2/E]$

1. Put the observed frequencies in L1 and the expected proportions in L3.

2. Highlight L2 as at the top of screen (1), and type **L3*** sum(**L1** as at the bottom of the screen (sum under ↑**LIST<MATH>5**).

3. Press **ENTER** for the expected frequencies (E) in L2, as in screen (2). (Of course, if these values are given or if they are easy to calculate without the spreadsheet, just enter them into L2.)

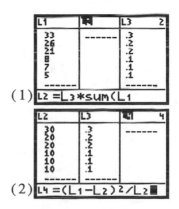

4. Highlight L4 and type $(L1-L2)^2/L2$ (as in the last line of screen (2)), and then press **ENTER** for the chi-square contribution of each color in L4, as in screen (3). The last row (color blue) makes the largest contribution, 2.5.

5. Press ↑**QUIT** to return to the Home screen.

Sum the above contributions with sum(L4, and then press **ENTER** for the chi-square statistic, which is 5.95 (see screen (4)).

P-Value [pg. 551]

6. Press ↑**DISTR** 7:χ^2cdf(**5.95,E99,5** and then **ENTER** for a p-value = 0.311 df = 5, or 6 − 1 (see screen (5)).

7. Since the p-value = 0.311 is larger than the significance level α = 0.05, we would fail to reject Mars's claim.

To graphically compare the observed frequencies with the expected frequencies, put the integers 1 to 6 in L6 as in screen (6).

Note: The observed proportions could be calculated in L5 by highlighting L5, typing L1/sum(L1), then pressing **ENTER**. (The first value in L5, or .33, compares with the the first expected proportion in L3 (screen (3) or .30).

Set up Plot1 and Plot2 as in screen (7) with different marks (a square for observed values and a dot for expected). Press **ZOOM** 9:ZoomStat, and then press **TRACE** and ▶ a few times for screen (8).

Notice that the points for each color are relatively close together. The last color (blue) has the greatest discrepancy, but this was found not significant above.

Note: The observed proportions in L5 and the expected proportions in L3 could have been plotted (see Figure 10.5 of text [pg. 551]), but the graph would look the same—only the y-axis scale would change.

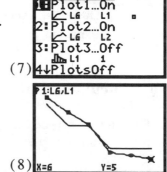

CONTINGENCY TABLES [pg. 557]

TABLE 10-1 Chapter Problem [pg. 541]: The table below is from a study that compared noncombat mortality rates for U.S. military personnel who were deployed in combat situations to those not deployed. Is there a relationship between cause of death and whether military personnel were deployed to a combat zone?

	Unintentional injury	Illness	Homicide or suicide
Deployed	183	30	11
Not deployed	784	264	308

EXAMPLE [pg. 560]: Find the expected frequencies for the lower-left cell of the above problem assuming independence between cause of death and whether the person was deployed.

E = (row total)*(column total)/(grand total)
  = (784+264+308)*(183+784)/(183+30+11+784+264+308) = 829.9
  = as in screen (9)

(9)

Note: This method is not efficient, but the output of STAT<TESTS> C:χ2-Test below gives all of the expected values automatically (see screen (15)).

EXAMPLE [pg. 561]: At the 0.05 significance level, use the data in the table above to test the claim that the cause of a noncombat death is independent of whether the military person was deployed in a combat zone.

Note: On a TI-82, calculate all the expected values and then calculate the chi-square test statistic using the spreadsheet as in the first example of this chapter.

1. Enter the data in a matrix. (Use matrix [D] as the Data matrix, but any matrix will do.) Press **MATRX<EDIT> 4**:[D] **2 ENTER 3 ENTER** for screen (10). You might have values other than zero in your matrix, but don't bother to change them because you will be entering values over them.

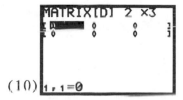

(10)

   Note. If the size of the previous matrix had more than one digit for the number of rows or columns, you will have to delete the second digit with **DEL**.

Type 183 **ENTER** 30 **ENTER** 11 **ENTER** 784 **ENTER** 264 **ENTER** 308 **ENTER** for screen (11).

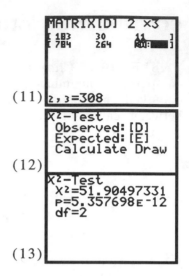

H_0: The cause of death is independent of whether the person was deployed in a combat zone.

(11)

2. Press **STAT<TESTS> C:**χ^2**-Test** and have the resulting screen look like screen (12) with [D] pasted by pressing **MATRX<NAMES> 4:**[D] and [E] pasted with **MATRX<NAMES> 5:**[E].

(12)

3. Highlight Calculate in the last line of screen (12), and then press **ENTER** for screen (13) with a test statistic of 51.9 and a p-value = 0.00000000000536.

(13)

4. Since the p-value = 0.000$^+$ is less than the significance level 0.05, we reject the null hypothesis. It appears that whether or not a person is deployed to a combat zone does seem to be related to the cause of death.

To compare the observed and expected frequencies on the same screen, press **MODE** and make the resulting screen look like screen (14) by highlighting **0** decimals in the second row and pressing **ENTER**. Press ↑**QUIT** to return to the Home screen.

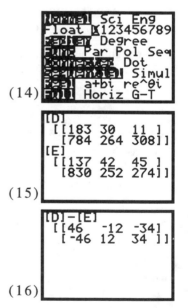

(14)

Note: Do not forget to change the **MODE** back to Float decimal when you have finished the following.

Paste [D] to the Home screen, and press **ENTER**. Do the same thing for matrix [E] for screen (15).

(15)

Paste [D], type −, and then paste [E] and press **ENTER** for the differences between the observed and expected values in screen (16). Notice that the largest relative difference is the last value in the first row with a chi-square contribution of $(O - E)^2/E = (-34)^2/45 = 26$. The next-largest contribution is the first value of the first row, or $46^2/137 = 15$. All other contributions are less than 4 to zero decimals.

(16)

Those that were deployed to a combat zone seemed to have more unintended injuries but less homicides or suicides than would be expected if there were independence between type of death and deployment to a combat zone.

11

Analysis of Variance

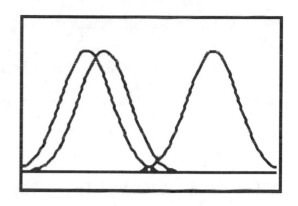

In this chapter you will learn about the **STAT<TEST>F:ANOVA** function for doing one-way ANOVA problems. You will also be introduced to the program A1ANOVA, which will extend your ability to do two-way ANOVA for both two-factor designs (with an equal number of observations in each cell), and randomized block designs. The purpose is to duplicate the Minitab ANOVA tables in the text with the TI-83. Follow the text for the proper interpretation of these tables.

ONE-WAY ANOVA [pg. 582]

The following procedure works for both equal and unequal sample sizes.

EXAMPLE [pg. 583]: Given the sample data below of the time intervals (in minutes) between eruptions of the Old Faithful geyser in Yellowstone National Park, use the TI-83 calculator to test the claim that the four samples come from populations with the same mean (H_0: $\mu_{1951} = \mu_{1985} = \mu_{1995} = \mu_{1996}$).

1951 (L1):	74	60	74	42	74	52	65	68	62	66	62	60
1985 (L2):	89	90	60	65	82	84	54	85	58	79	57	88
1995 (L3):	86	86	62	104	62	95	79	62	94	79	86	85
1996 (L4):	88	86	85	89	83	85	91	68	91	56	89	94

1. Put the data in L1 to L4 as indicated in the above table.   (1)

2. Press **STAT<TESTS>F:ANOVA** for **ANOVA(** on the Home screen. Type L1,L2,L3,L4 for screen (1).

3. Press **ENTER** for screens (2) and (3), with the test statistic F = 6.897 = 6.90 and a p-value = 0.0006604 = 0.001 in screen (2), and the pooled standard deviation, or Sxp = 12.22, in screen (3). The results will be similar to those in the Minitab table in the text [pg. 584].   (2)

Note: Program A1ANOVA, which is introduced in the next section,   (3)

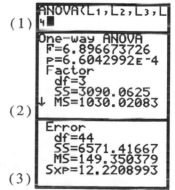

gives the means and standard deviations of the raw data stored in the list and also gives the 95% confidence intervals similar to Minitab output. The program accepts sample summary statistics (means, standard deviations, and sample sizes) as an input option in addition to the raw data option.

TWO-WAY ANOVA [pg. 596]

To better understand how to use program A1ANOVA for two-way ANOVA, study the following example. It uses the program to do the introductory example of the section, which is a one-way ANOVA. (The availability of program A1ANOVA is given in the Appendix. Be sure your copy is protected as explained by the warning on page 115.)

TABLE 11-3 [pg. 597]: Are bad movies longer than good movies, or does it just seem that way? Refer to the sample data below. An examination of the summary statistics seems to suggest that the mean lengths of movies do differ, with movies rated as excellent tending to be longer. But are those differences significant? Test the claim that the three categories of movies have the same mean length.

Lengths (in min) of Movies Categorized by Star Ratings		
Fair (1) 2.0-2.5 Stars	Good (2) 3.0-3.5 Stars	Excellent (3) 4.0 Stars
98	93	103
100	94	193
123	94	168
92	105	88
99	111	121
110	115	72
114	133	120
96	106	106
101	93	104
155	129	159

The data in the table above is going to be stored in matrix [D], with 30 rows and 2 columns as shown at the right.

98	1
100	1
123	1
92	1
99	1
110	1
114	1
96	1
101	1
155	1
93	2
94	2
94	2
105	2
111	2
115	2
133	2
106	2
93	2
129	2
103	3
193	3
168	3
88	3
121	3
72	3
120	3
106	3
104	3
159	3

1. **Enter Data into Matrix [D].**

   (a) Press **MATRX**<**EDIT**>**4**:[D] for the Edit screen of matrix
   [D]. Type 30 **ENTER** 2 **ENTER** for 30 rows and 2 columns,
   and the cursor will be at the first slot of the matrix as
   in screen (4). If your screen does not have all zeros,
   don't worry: you will be typing over these values.

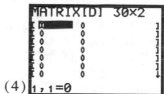

(4)

   (b) You can enter data by column by typing 9 8 and then
   pressing ▼ 100 ▼ 123 ▼ ...159 **ENTER** for all 30 movie
   lengths in minutes, with **ENTER** moving the cursor to
   the bottom of the second column. Then press 3 ▲ 3 ▲
   ...2 ▲ 2 ▲ ...1 ▲ 1 ▲ for ten 3's, ten 2's, and ten 1's for the
   two columns of values, as given on the preceding page.
   (You could also have entered the data row by row, as
   98 **ENTER** 1 **ENTER** 100 **ENTER** 1 **ENTER** ...104 **ENTER** 3 **ENTER**
   159 **ENTER** 3 **ENTER**.)

(5)

2. **Running Program A1ANOVA.**

   (a) Press **PRGM**<**EXEC**> and highlight program
   A1ANOVA. Press **ENTER**, and prgmA1ANOVA is pasted
   to the Home screen as in screen (5).

(6)

   (b) Press **ENTER** for option screen (6). Press **1** for a one-
   way ANOVA design and the instructional screen (7),
   which reminds you how to store the data in matrix
   [D].

(7)

   (c) Press **ENTER** for screen (8), which gives you three
   options: use data stored in matrix [D], or input
   summary statistics from screen prompts, or quit
   if the data is not ready.

(8)

   (d) Press **1**, since the data is in [D], for the ANOVA
   table of screen (9) with F = 1.17, a p-value = 0.325,
   and a pooled standard deviation of SP = 25.99, all of
   which agree with the Minitab display in the text
   [pg. 596].

(9)

   (e) Pressing **ENTER** gives screen (10), with the sample
   sizes, mean, and standard deviation for each sample
   given. (Use the ▶ key to reveal the complete
   standard deviations.) Pressing **ENTER** again gives
   screen (11) with the 95% confidence intervals based
   on the pooled standard deviation. (All of the above
   are similar to the Minitab display in the text
   [pg. 596]).

(10)

(11)

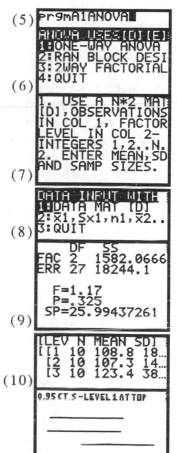

Note: Although screen (7) does not give Mean Squares (MS), they are easily calculated by dividing Sum of Squares (SS) by Degrees of Freedom (DF).

You will be using the data currently in matrix [D] in the next section, so you might want to save it now in another matrix, such as [C], as follows:

Press **MATRX 4**:[D], and [D] is pasted to the Home screen.

Press **STO►** and then **MATRX 3**:[C] for [D] → [C] on the Home screen.

Press **ENTER**, and the matrix is copied into [C] and shows on the Home screen.

Two-Factor Design with an Equal Number of Observations per Cell

TABLE 11-4 [pg. 598]: This example is a continuation of the previous example, so if you have been using matrix [D] for other problems, you can now return the data (e.g., [C] → [D]). A second factor, not given before, is the Motion Picture Association of America (MPAA) ratings, which are shown in the table below. For each star rating category in the previous example, the data were arranged so that the first five movies had a MPAA rating of G, PG, or PG-13, and the last five were all rated R. The above ratings give six cells, or groups of data, of five values each; or you could say there are five observations for each cell, or equal replicates in the language of the program **A1ANOVA 3:2WAY FACTORIAL** that you will be using.

The data for this example are repeated below, and the second factor is now identified. Also shown are the contents of matrix [D], now 30 **✕** 3 with the second (B) factor identification added in the third column at the right.

Lengths (in min) of Movies Categorized by Star Ratings and MPAA Ratings			
	Fair A (1) 2.0–2.5 Stars	Good A (2) 3.0–3.5 Stars	Excellent A (3) 4.0 Stars
MPAA Rating G/PG/PG-13 B (1)	98 100 123 92 99	93 94 94 105 111	103 193 168 88 121
MPAA Rating R B (2)	110 114 96 101 155	115 133 106 93 129	72 120 106 104 159

98	1	1
100	1	1
123	1	1
92	1	1
99	1	1
110	1	2
114	1	2
96	1	2
101	1	2
155	1	2
93	2	1
94	2	1
94	2	1
105	2	1
111	2	1
115	2	2
133	2	2
106	2	2
93	2	2
129	2	2
103	3	1
193	3	1
168	3	1
88	3	1
121	3	1
72	3	2
120	3	2
106	3	2
104	3	2
159	3	2

1. Enter Data into Matrix [D].

Press **MATRX** [EDIT] **4**:[D] for the Edit screen of matrix [D]. Type 30 **ENTER** 3 **ENTER** for 30 rows and 3 columns, and the cursor will be at the first slot of the matrix. If you do not already have the first two columns from before, enter them as explained in the last example:

(a) Using the first method of the preceding example, for the first two columns, use ► to go to the first element of the third column, type 1, and then press ▼ 1 ▼ 1 ▼ 1 ▼ 1 ▼ for five 1's in the third column. Follow this by five 2's, five 1's, . . ., 2 ▼ 2 ▼ **ENTER**, as in the table at the far right of page 96. The final **ENTER** is to store the last 2.

(b) For the second method of the preceding example, you could enter all the data row by row. Type 98 **ENTER** 1 **ENTER** 1 **ENTER** 100 **ENTER** 1 **ENTER** 1 **ENTER** . . .159 **ENTER** 3 **ENTER** 2 **ENTER**.

Note: If the first two columns already exist, use (a) to enter the last column. The second column is the integer value for factor A (star rating); the third column gives the integer value of factor B (MPAA rating) for each movie whose length, in minutes, is given in the first column. This order of the data differs from the Minitab example in the text [pg. 599].

2. Run Program A1ANOVA 3:2WAY FACTORIAL.

Run program A1ANOVA as you did in the last example, but when you get to screen (12), press 3 for screen (13) and continue for the output in screens (14) and (15). (12)

Note: A busy signal shows in the upper-right-hand corner of the display screen while the calculations are being done.

The ANOVA table output in screens (14) and (15) is similar to the Minitab display in the text [pg. 599]. But here we have the star rating first (A) and the MPAA rating second (B); in the Minitab output, the order is reversed. You can obtain the MS values from the SS and DF (e.g., MS ERR= SS ERR÷DF ERR, or 15956÷24 = 664.833 = 665).

(13)

(14)

(15)

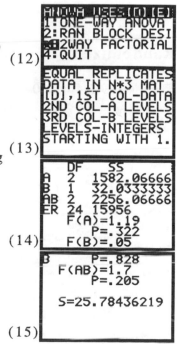

Special Case: One Observation per Cell and No Interaction [pg. 602]

EXAMPLE [pg. 602] Suppose you have only the first score in each cell of the preceding example, as in the following table, and in matrix [D], as in screen (16).

Lengths (in min) of Movies Categorized by Star Ratings and MPAA Ratings			
	Fair A (1) 2.0–2.5 Stars	Good A (2) 3.0–3.5 Stars	Excellent A (3) 4.0 Stars
G/PG/PG-13 B (1)	98	93	103
R B (2)	110	115	72

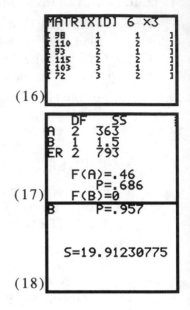

(16)

(17)

(18)

Running program **A1ANOVA 3:2WAY FACTORIAL** gives screens (17) and (18) of output, which are similar to the Minitab display in the text except for Minitab's rounding [pg. 603].

Randomized Block Design [pg. 603]

EXAMPLE [pg. 604]: Refer to the sample data below. Test the claim that the grade of gasoline does not affect mileage, blocking on the four cars used.

Mileage (mi/gal) for Different Grades of Gasoline (Treatment)			
B	Regular	Extra	Premium
L Car1	19	19	22
O Car2	33	34	39
C Car3	23	26	26
K Car4	27	29	34
mean	25.5	27	30.25
S.D.	5.97	6.27	7.68

19	1	1
33	1	2
23	1	3
27	1	4
19	2	1
34	2	2
26	2	3
29	2	4
22	3	1
39	3	2
26	3	3
34	3	4

Notice that the table has been turned around from the text to keep the treatment of primary interest (grades of gasoline) in columns, as we have been doing.

The contents of a 12 × 3 matrix [D] are also given above right, with the treatment integer in the second column and the blocking integer (car) in the third column. Refer to the previous examples to find out how to enter the data into matrix [D].

Run program A1ANOVA 2:RAN BLOCK DESI. The ANOVA table output is shown in screens (21) and (22) and is similar to the Minitab display in the text [pg. 605].

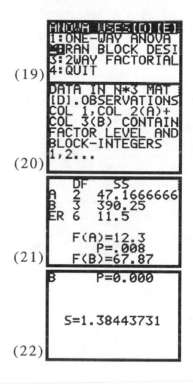

(19)

(20)

(21)

(22)

12

Statistical Process Control

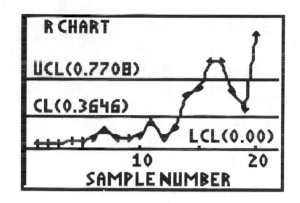

In this chapter you will learn how to plot run charts and control charts for range, $\bar{x}$, and proportions.

RUN CHARTS [pg. 621]

EXAMPLE [pg. 621]: Treating the 175 axial loads on aluminum cans (given and stored as LALCAN on page 21 with mean $\bar{x} = 267.1$ as on page 27) as a string of consecutive measurements, construct a run chart by using a vertical axis for the axial loads and a horizontal axis to identify the order of the sample data.

1. Generate integers from 1 to 175 and store them in L1 with ↑LIST<OPS>5:seq(X,X,1,175,1) **STO►** L1. Then press **ENTER** (see screen (1)).

   Note: The TI-83 can only go from 1 to 99.

2. Set up Plot1 as an xy-Line plot as in screen (2). (All other Stat Plots should be off.) Paste **ALCAN** as on page 12 (step 6).

   Note: TI-83 users can put the first 99 measurements in L2 instead of LALCAN.

3. Press the Y= key and let Y1= 267.1 (the mean axial load) as in screen (3). (All other Y – plots should be off.)

4. Press **ZOOM 9**:ZoomStat, and adjust the **WINDOW** to better fill the screen by changing Xmin = 0 and Xmax = 175. Then press **TRACE** for screen (4), which is similar to the plot in the text [pg. 621]).

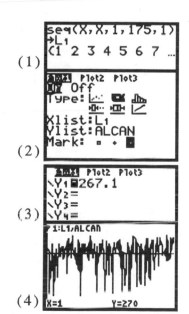

(1)

(2)

(3)

(4)

CONTROL CHART FOR MONITORING VARIATION: THE R CHART [pg. 625]

EXAMPLE [pg. 626]: Refer to the ranges of the axial loads of aluminum cans (of samples of size 7 collected each day for 25 working days) given in Table 12-1 of the text and repeated below. Construct a control chart for R.

78 77 31 50 33 38 84 21 38 77 26 78 78 17 83 66 72 79 61 74 64 51 26 41 31

1. Put the day in L1 (integers from 1 to 25) and the corresponding ranges from above in L2. Use **STAT<CALC>1: 1-VarStats L2** for $\bar{R}$ = 54.96, from which you can calculate the control limits as in the text [pg. 627] with UCL = 105.74, CL = 54.96, and LCL = 4.18.

2. Set up Plot1 as in screen (5).

3 Set up the Y = editor with the control limits as in screen (6).

4. Press **ZOOM 9**:ZoomStat, and adjust the **WINDOW** to include the control limits by changing Ymin = ‾25 and Ymax = 125. Then press **TRACE** for screen (7), which is similar to the R-chart in the text [pg. 628].

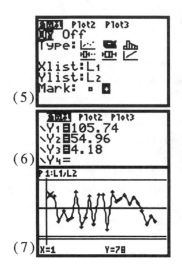

(5)

(6)

(7)

MONITORING PROCESS MEAN: CONTROL CHART FOR $\bar{x}$ [pg. 630]

EXAMPLE [pg. 630]: Refer to the means of the axial loads of aluminum cans (of samples of size 7 collected each day for 25 working days) given in Table 12-1 of the text and repeated below. Construct a control chart for $\bar{x}$. Use the control limits as calculated in the text [pg. 631]: UCL = 290.2, CL = 267.1, and LCL = 244.1.

252.7 247.9 270.3 267.0 281.6 269.9 257.7 272.9 273.7 259.1 275.6 262.4 256.0
277.6 264.3 260.1 254.7 278.1 259.7 269.4 266.6 270.9 281.0 271.4 277.3

Put the day in L1 (integers from 1 to 25) and the corresponding means from above in L2. Set up Plot1 as in screen (5) and the **Y=** editor as in screen (8).
Press **ZOOM 9**:ZoomStat, and adjust the **WINDOW** to include the control limits by changing Ymin = 235 and Ymax = 300. Then press **TRACE** for screen (9), which is similar to the $\bar{x}$-chart in the text [pg. 631].

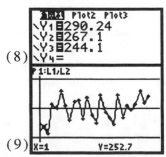

(8)

(9)

CONTROL CHART FOR PROPORTION P [pg. 637]

EXAMPLE [pg. 637]: In each of 13 consecutive and recent years, 100,000 subjects were randomly selected and the number who died from respiratory tract infections was recorded, with the results given below. Construct a control chart for p with the control limits as calculated in the text [pg. 638]: UCL = 0.000449, CL = 0.000288, and LCL = 0.000127.

$$25 \quad 24 \quad 22 \quad 25 \quad 27 \quad 30 \quad 31 \quad 30 \quad 33 \quad 32 \quad 33 \quad 32 \quad 31$$

Put the year in L1 (integers from 1 to 13) and the corresponding number of deaths in L2. From the Home screen, let L2÷100000 **STO►**L2 to store the proportion of deaths in L2.

(10)

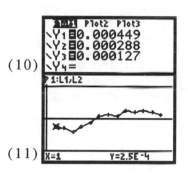

Set up Plot1 as in screen (5) and the **Y=** editor as in screen (10). Press **ZOOM** 9:ZoomStat, and adjust the **WINDOW** to include the control limits by changing Ymin = 0.00009 and Ymax = 0.0005. Then press **TRACE** for screen (11), which is similar to the p-chart in the text [pg. 639].

(11)

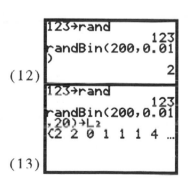

COMPUTER PROJECT [pg. 644]

Simulate 20 days of manufacturing heart pacemakers with a 1% rate of defective units. Use the TI-83 to simulate taking a sample of 200 pacemakers each day.

One sample can be generated as in screen (12), with rand and randBin both from **MATH<PRB>** and 2 defective as in the last line of the screen.

Twenty samples can be generated and stored in L2 as in screen (13), where the seed had been reset in the first two lines. (Be patient and notice the busy symbol in the upper-right hand corner while the list is being generated.) These samples vary from 0 to 4 defective. The complete list of defective units for the 20 samples can be seen using **►** or looking in L2.

(12)

(13)

13

Nonparametric Statistics

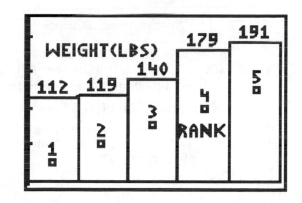

In this chapter you will perform most calculations using the TI-83's spreadsheet and data-sorting features. Examples are given for all the nonparametric tests in the text. When sample sizes are too large for the tables of critical values given in the text, a test statistic, z, which is normally distributed can be calculated for many methods. You can find the p-value for the calculated z by using the normalcdf function as shown in screen (3) on page 106.

SIGN TEST [pg. 654]

Claims Involving Two Dependent Samples [pg. 655]

EXAMPLE [pg. 655]: The following data was obtained when 14 subjects were tested for reaction times of their right and left hands. (Only right-handed subjects were used.) Test the claim of no difference between the right- and left-hand reaction times at the 0.05 level of significance. (Reaction times are in thousandths of a second.)

Subject	1	2	3	4	5	6	7	8	9	10	11	12	13	14
Right hand	191	97	116	165	116	129	171	155	112	102	188	158	121	133
Left hand	224	171	191	207	196	165	177	165	140	188	155	219	177	174
Sign of difference	−	−	−	−	−	−	−	−	−	−	+	−	−	−

If the hand used had no effect on reaction time, we would expect about half the subjects to be faster with their left hand and half to be faster with their right hand. Of the 14 subjects, however, only 1 has a positive sign and 13 have negative signs for the differences in reaction time, unlike the expected (the average in the long run) 7 of each. This is a binomial distribution with n = 14 and p = 0.5 for a mean of n*p = 14*0.5 = 7.

Note: You might want to review the binomial distribution in Chapter 4.

You want the probability in the tail or the probability of getting 1 or fewer positives differences or $P(0) + P(1)$.

Press ↑**DISTR A:binomcdf(14,.5,1** and then **ENTER** for 0.0009155 as in screen (1). Since this is a two-tail test, multiply this value by 2 for a p-value = 0.00183, as in the bottom line of screen (1).

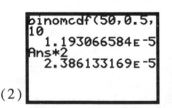

(1)

Since the p-value is less than the significance level of 0.05, we reject the null hypothesis that the hand used makes no difference in reaction time and conclude that the left hand is significantly slower in reaction time than the right. In the sample, only once in 14 times was the left hand faster than the right.

Note: If the sample sizes become very large, we could use the normal distribution to approximate the binomial (see Chapter 5 and the next example).

Claims Involving Nominal Data [pg. 659]

EXAMPLE [pg. 659]**:** A company acknowledges that about half the applicants for new jobs were men and half were women. All applicants met the basic job-qualification standards. Test the null hypothesis that men and women are hired equally by this company if of its last 50 hires only 10 were women.

H_0: $p = 0.5$ H_1: $p \neq 0.5$

This is a binomial distribution with $n = 50$ and $p = 0.5$, and you want to find the probability of having 10 or fewer women, or $P(0) + P(1) + P(2) + ... + P(10)$.

Press ↑**DISTR A:binomcdf(50,0.5,10** and then **ENTER** for 0.00001193 as in screen (2). Since this is a two-tail test, multiply this value by 2 for a p-value of 0.00002386, as in the bottom line of screen (2). Since the p-value is so small, there is sufficient evidence to warrant rejection of the claim that the hiring practices are fair.

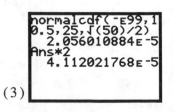

(2)

Note: If the sample sizes become larger than 999, we could use the normal distribution to approximate the binomial (see Chapter 5). Since $n*p = 50*0.5 = 25 > 5$ and $n*q = 25 > 5$, we can use the normal approximation for the above problem and will do so to review the procedure: $\mu = n*p = 25$ and $\sigma = \sqrt{(n*p*q)} = \sqrt{(50*.5*.5)} = \sqrt{(50)} \div 2$. We must find the area under the normal curve from the far left to 10.5 to include all of $P(10)$ of the binomial distribution.

(3)

Press ↑DISTR 2:normalcdf(-E99,10.5,25,√(50)÷2) then press ENTER for 0.0000206 as in screen (3). Since this is a two tail test multiply this by 2 for a p-value = 0.0000411 as in the bottom line of screen (3). This is still a very small p-value (but not as accurate as above) so we would come to the same conclusion and reject the null hypothesis.

Claims About the Median of a Single Population [pg. 661]

EXAMPLE [pg. 661]: Use the sign test to test the claim that the median value of the 106 body temperatures of healthy adults (used in Chapter 7) is less than 98.6°F. The data set has 68 subjects with temperatures below 98.6°F, 23 subjects with temperatures above 98.6°F, and 15 subjects with temperatures equal to 98.6°F. (The steps below show how you can find these values.)

Discounting the 15 temperatures equal to 98.6°F because they do not add any information to this problem, n = 68 + 23 = 91. If the median were 98.6, we would expect about half of these 91 values to be below the median (negatives) and half to be above the median (positives). This is a binomial distribution with $\mu = n*p = 45.5$ and $\sigma = \sqrt{(n*p*q)} = \sqrt{(91*(1/2)*(1/2))} = \sqrt{(91)}/2$. We want the probability of getting 23 or fewer positive values.

```
binomcdf(91,0.5,
23)
        1.261076831E-6
```
(4)

Press ↑DISTR A:binomcdf(91,0.5,23) and then press ENTER for a p-value = 0.00000126 as in screen (4).

With such a small p-value, there is good evidence that in fact there are too few positive and significantly more negative values than would be expected. The data supports the claim that the median body temperature of healthy adults is less than 98.6°F.

If the 106 body temperatures are saved in a list (e.g., LBTEMP), the following steps show how the numbers below and above the median given could be obtained.

1. Press STAT 5: SetUpEditor LBTEMP,L1,L2 ENTER, and then STAT 1:Edit for the results in screen (5).

2. Highlight L1 at the top of screen (5), paste LBTEMP, and subtract 98.6 as in the last line. Press ENTER for screen (6). Since the first two values of LBTEMP are 98.6, the difference in L1 is 0. The next two values are 0.6 less than 98.6, and the fifth value is 99, or 0.4 greater than 98.6 and so on.

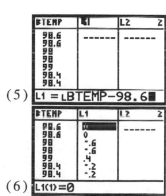
(5)
(6)

3   Use **STAT 2**:SortA(**L1** and then press **ENTER** to sort the differences in L1 from low negative values to high positive values.

4.   Using ▼, observe that <u>68</u> values are below 98.6 (negative), as in screen (7). Zero differences go from row 69 to row 83, or <u>15</u> values are equal to 98.6. Positive differences, or values above 98.6, go from row 84 to row 106, or <u>23</u> values. These results are the same as those given above.

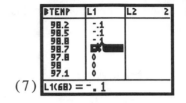

(7) L1(6B) = -.1

WILCOXON SIGNED-RANKS TEST FOR TWO DEPENDENT SAMPLES [pg. 665]

EXAMPLE [pg. 666]: The following data (used in the first example of this chapter) was obtained when 14 subjects were tested for reaction times with their right and left hands. (Only right-handed subjects were used.) Use the Wilcoxon signed-ranks test to test the claim of no difference between reaction times with right and left hands. Use a significance level of $\alpha = 0.05$.

Subject	1	2	3	4	5	6	7	8	9	10	11	12	13	14
Right hand (L1)	191	97	116	165	116	129	171	155	112	102	188	158	121	133
Left hand (L2)	224	171	191	207	196	165	177	165	140	188	155	219	177	174

The following steps show a way of finding the sum of the positive ranks and the sum of the absolute values of the negative ranks.

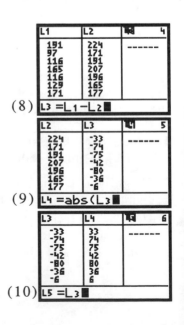

1.   Put the right-hand times in L1 and the left-hand times in L2. Highlight L3 and then enter L1-L2 for the bottom line of screen (8). Press **ENTER** to calculate the difference, as in screen (9).

(8) L3 =L1-L2

2.   Highlight L4 and enter abs(L3, as in the bottom line of screen (9). Press **ENTER** for all positive values, as in screen (10).

Note: abs(from **MATH<NUM>1**.

(9) L4 =abs(L3

3.   Copy the differences in L3 into L5 by highlighting L5, entering L3 (as in the bottom line of screen (10)), and then press **ENTER** for screen (11).

(10) L5 =L3

4. Press **STAT 2**:Sort**A(L4,L5 ENTER,** for Done as in screen (12), which puts L4 in order and carries along L5. Return to the **STAT** editor for screen (13).

5. Highlight L5 and enter L5 ÷ L4, as in the last line of screen (13). Pressing **ENTER** gives you a column of positive and negative 1's in L5, as in screen (14).

6. Generate the integers from 1 to 14 in L6 with ↑**LIST<OPS> 5:seq(X,X,1,14,1** as in the bottom line of screen (14). Press **ENTER** and modify L6 so that the ranks of the values in each row of L4 are given in L6 (see screen (15)). Since the fourth and fifth values in L4 are both 33, the 4 and 5 in L6 were both changed to their average, or 4.5. (But note that 4.5 + 4.5 = 4 + 5.)

7. Engage ↑**QUIT** to return to the Home screen. Type **14(14+1)÷2 STO► A ENTER** for the sum of the ranks (or 105) to be stored as A. Engage ↑**LIST<MATH>5:sum(L6** for the sum of the values in L6, which should also be 105. If you get a different answer, check your rankings (see screen (16)).

8. Now have L5 *L6 **STO► L6 ENTER,** which puts a sign on the ranks in L6. In this example, the first three are negative, and the fourth is positive; the rest are negative (see the second line of screen (17)).

9. (sum(L6) +A)÷2 **ENTER** for 4.5 or the sum of the ranks for the positive differences. Of course, this problem has just one value. Type **A - ↑ANS ENTER** for 100.5, the sum of the ranks for the negative differences.

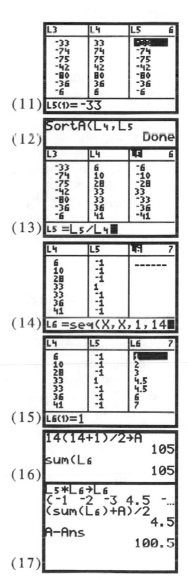

(11) L5(1)= -33

(12)

(13) L5 =L5/L4

(14) L6 =seq(X,X,1,14

(15) L6(1)=1

(16)

(17)

Since 4.5 is the smaller of the two and n = 14 ≤ 30, we find that the critical value in text Table A 8 is 21; and since the test statistic of 4.5 is less than 21, we reject the null hypothesis and conclude that there is a significant difference between right- and left-hand reaction times.

Because negative signs (there are 13) result from lower values with the right hand, it appears that right-handed people have faster reaction times with their right hand compared to their left hand.

WILCOXON RANK-SUM TEST FOR TWO INDEPENDENT SAMPLES [pg. 671]

EXAMPLE [pg. 673]: Samples of M&M® plain candies are randomly selected, and the red and yellow M&M®s are weighed, with the results listed in the table at the right. At the 0.05 level of significance, test the claim that weights of red M&M®s and yellow M&M®s have the same distribution.

The text bases its results on the sum of the ranks of the red M&M®s, which is R = 469.5. You will see how to calculate this in the numbered steps below. Since $n_1 = 21$ and $n_2 = 26$ are both greater than 10, you can use a normal distribution with $\mu_r = n_1(n_1 + n_2 + 1)/2$.

Or type **21(21+26+1)/2 STO▶ M** and then press **ENTER** for 504, as in the top lines of screen (18).

$\sigma_r = \sqrt{(n_1 \cdot n_2 \cdot (n_1 + n_2 + 1)/12)}$, or type

√ (21*26*(21+26+1)/12 STO▶ S

and then press **ENTER** for 46.7333.   (18)

$z = (R - \mu_r)/\sigma_r = (469.5 - 504)/46.733 = {}^-0.7382.$

Find the area in the left tail of the normal distribution using **↑DISTR 2:normalcdf(-E99,-.738**. Then press **ENTER** for a p-value = 0.230257*2 = 0.4605, as in screen (19). Thus we fail to reject the null hypothesis of similar distributions because the p-value is so large.   (19)

Note: The mean rank for red M&M®s is 469.5/21 = 22.36, and the mean rank for yellow M&M®s is 658.5/26 = 25.33. These means are fairly close, so we are not surprised that we fail to reject the null hypothesis.

The following steps give a method for finding the needed   ranks.

1. The weights, in grams, are given in the first
   column of the table on the right. Put these values
   in L1. The 21 weights of the red M&M®s are listed
   first and are identified by a 1 in the second
   column. The other 26 weights are yellow M&M®s
   and are identified by a 0.
   Put 21 1's followed by 26 0's in L2.

2. Make a copy of L1 into L3. Make a copy of L2 into L5.

L1	L2
0.87	1
0.933	1
0.952	1
0.908	1
0.911	1
0.908	1
0.913	1
0.983	1
0.92	1
0.936	1
0.891	1
0.924	1
0.874	1
0.908	1
0.924	1
0.897	1
0.912	1
0.888	1
0.872	1
0.898	1
0.882	1
0.906	0
0.978	0
0.926	0
0.868	0
0.876	0
0.968	0
0.921	0
0.893	0
0.939	0
0.886	0
0.924	0
0.91	0
0.877	0
0.879	0
0.941	0
0.879	0
0.94	0
0.96	0
0.989	0
0.9	0
0.917	0
0.911	0
0.892	0
0.886	0
0.949	0
0.934	0

Screen (18):
```
21(21+26+1)/2→M
                504
√(21*26*(21+26+1
)/12)→S
        46.73328578
(469.5-M)/S
        -.7382318496
```

Screen (19):
```
normalcdf(-E99,-
.738
         .2302571507
Ans*2
         .4605143014
```

3. Press **STAT 2:SortA(L3, L5 ENTER** to sort the values in L3 and carry along the values in L5.

4. Generate the integers from 1 to 47 in L4 with ↑**LIST<OPS> 5:seq(X,X,1,47,1** and then modify L4 so that it has the ranks of the values in L3 in L4 (see screen (20)). **(20)**

Note: Notice how the tie in the 7th and 8th slots was handled (both given a rank of 7.5) in screen (20). The 10th and 11th values are also the same and get the ranks of 10.5. The 20th, 21st, and 22nd items, which are the same, all get a rank of 21. The 24th and 25th values both get 24.5; the 31st, 32nd, and 33rd values all get ranked 32. **(21)**

5. After entering all the ranks, press ↑**QUIT** to return to the Home screen. Type 47(47+1)÷2 **STO▸ A** (for 1128, the sum of the integers from 1 to 47 stored in **A**). Then press **sum(L4) ENTER** as a check on your work (see screen (21)). **(22)**

6. Multiply the 1's and 0's in L5 by the ranks in L4 and put the results in L6 (**L4✱ L5 STO▸ L6**).

Note: This will copy all the ranks for the red M&M® sample, but the yellow M&M® ranks will be zeroed out.

Then **sum(L6 ENTER** gives the sum of the red M&M® ranks, or 469.5 = R1 of the text, as shown in screen (22).

Subtracting 469.5 from the total sum of all the ranks gives us the sum of the ranks for the yellow M&M®s, or **A−↑ANS ENTER** for 658.5 = R2 of the text.

KRUSKAL–WALLIS TEST [pg. 678]

EXAMPLE [pg. 680]: The data in the table at the right on page 112 are the mean time intervals between eruptions of the Old Faithful geyser in Yellowstone National Park for four different years. (This is the chapter problem of Chapter 11.) Put the time intervals (in minutes) in L1 and the four years designated by 1, 2, 3, or 4 in L2. Use the Kruskal–Wallis test to test the null hypothesis that the different years have time intervals with identical populations.

The ranks of the time intervals for each year are shown in the text and are calculated in the numbered steps that follow. They are R1 = 157, R2 = 265.5, R3 = 358.5, and R4 = 395. Since $n_1 = n_2 = n_3 = n_4 = 12$ are each at least 5,

H (that follows) is chi-square distributed with $k - 1 = 4 - 1 = 3$ degrees of freedom. Below, $N = n_1 + n_2 + n_3 + n_4 = 12 + 12 + 12 + 12 = 48$.

$$H = 12/(N(N+1))(R_1^2/n_1 + R_2^2/n_2 + R_3^2/n_3 + R_4^2/n_4) - 3(N + 1)$$

$$= 12/(48*49)*(157^2/12 + 265.5^2/12 + 358.5^2/12 + 395^2/12) - 3(49)$$

$$= 14.431$$

Press ↑**DISTR** 7:χ^2cdf(**14.431,E99,3**) and then **ENTER** for the p-value = 0.00237 (see screen (23)). Since the p-value is less than 0.05, we reject the null hypothesis of identical populations.

Note: The average ranks from the foursamples show a trend of increasing time intervals between eruptions over the years.

(23)

Use the method described in the following steps to find the ranks of each of the four samples.

1. With the data in L1 and the year coded from 1 to 4 in L2 as in the table at the right, make a copy of L1 in L3 (with L1 **STO▶** L3 on the Home screen).

2. Press **STAT** 2:SortA(L3, L2 **ENTER** to sort the values in L3 and carry along the values in L2 (see L2 and L3 in screen (24) but notice that the cursor was moved to the 7th row).

(24)

3. Generate the integers from 1 to 48 in L4 with ↑**LIST**<OPS> 5:seq(**X,X,1,48,1 STO▶** L4 and then modify L4 so that it has the ranks of the values in L3 in L4 (see screen (24)). The 7th, 8th and 9th values are are all 60, so they are all given the same rank, 8. The 10th through the 14th value are all 62, so they are all given the average of 10, 11, 12, 13, 14, or 12. The 15th and 16th values are both 65, and are both ranked 15.5. There are two 18.5's, three 21's, three 24's, four 30.5's, four 34.5's, two 37.5s, three 40's, two 43.5's, and two 45.5's.

4. Screen (25) shows that the sum of the integers from 1 to 48 is 1176. As a check, the sum of the ranks in L4 is also 1176.

(25)

5. SortA(L2, L3,L4 to sort the values in L2, putting all the 1's first, followed by the 2's, and so on, and carry along the values in L3 and L4.

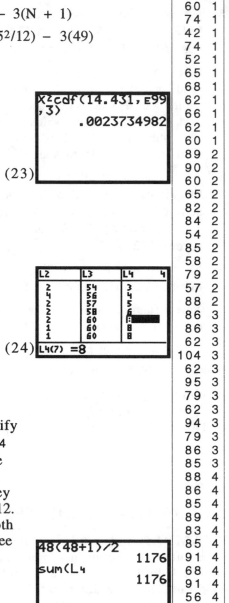

L1	L2
74	1
60	1
74	1
42	1
74	1
52	1
65	1
68	1
62	1
66	1
62	1
60	1
89	2
90	2
60	2
65	2
82	2
84	2
54	2
85	2
58	2
79	2
57	2
88	2
86	3
86	3
62	3
104	3
62	3
95	3
79	3
62	3
94	3
79	3
86	3
85	3
88	4
86	4
85	4
89	4
83	4
85	4
91	4
68	4
91	4
56	4
89	4
94	4

6. Generate 48 zeros in L5 with ↑LIST<OPS> 5:seq(0,X,1,48,1 STO► L5 and then replace the first 12 values with 1's, as shown in screen (26), which shows the 12th value L5(12) = 1 (as are all the values above it). The values below are all 0's.

7. Multiply L4 by L5 and store the results in L6. Only the first 12 values, or the ranks that go with 1951, will be in L6; all other rows will record a zero so that when you sum(L6 you get R1 = 157 (see screen (27)).

Repeat the above two steps, but this time put a 1 next to the 13th through 24th values. All other values are zero, so the sum will be R2 = 265.5. Similarly, R3 = 358.5 and R4 = 395.

(26)

L3	L4	L5	5
62	12	1	
62	12	1	
65	15.5	1	
42	1	▄	
90	42	0	
65	15.5	0	
85	30.5	0	

L5(12) =1

```
L4*L5→L6
{2 21 21 21 18. …
sum(L6
                157
```

(27)

RANK CORRELATION [pg. 685]

EXAMPLE [pg. 687]: *Business Week* magazine ranked business schools two different ways. Corporate rankings were based on surveys of corporate recruiters, and graduate rankings were based on surveys of MBA graduates. The table at the right is based on the results for 10 schools. Is there a correlation between the corporate rankings and the graduate rankings? The linear correlation coefficient r (section 9-2) should not be used because it requires normal distributions and the data consists of ranks, which are not normally distributed. Instead, the rank correlation coefficient should be used to test the claim that there is a relationship between corporate and graduate rankings (that is, $\rho_s \neq 0$).

Use a significance level of $\alpha = 0.05$.

School	Corp. Rank(L1)	Grad. Rank(L2)
PA	1	3
NW	2	5
Chi	4	4
Sfd	5	1
Hvd	3	10
MI	6	7
IN	8	6
Clb	7	8
UCLA	10	2
MIT	9	9

With the data in L1 and L2, the scatter diagram in screen (28) (as on page 25) shows little evidence of a relationship between the two variables.

Press **STAT**<TESTS>**E**:LinRegTTest and set up the screen for screen (29).

Highlight Calculate in the bottom line of screen (29), and then press **ENTER** for screens (30) and (31).

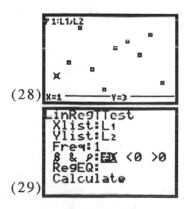

(28) X=1 ──── Y=3

```
LinRegTTest
Xlist:L1
Ylist:L2
Freq:1
β & ρ:≠0  <0  >0
RegEQ:
Calculate
```

(29)

The last line of screen (31) gives $r_s = 0.1030$.

Note: For the TI-82, press **STAT<CALC>5:LinReg(ax+b) L1,L2**
and then **ENTER** for r_s.

To test the hypothesis, use the critical value from Table A-9 of the text. Do not use Table A-6 (which was used for the Pearson correlation coefficient) because it requires the populations sampled to be normally distributed.   (30)

The critical value is $0.648 > 0.103$, so we fail to reject the null hypothesis. It appears that corporate recruiters and business school graduates have different perceptions of the qualities of the schools.   (31)

Note: The t and p-values of screen (30) are for the Pearson correlation coefficient and do not apply for the ranks.

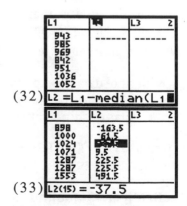

RUNS TEST FOR RANDOMNESS [pg. 696]

EXERCISE 10 [pg. 704]: Trends in business and economics applications are often analyzed with the runs test. Listed below (in order by row) are the annual high points of the Dow-Jones Industrial Average for a recent sequence of years. First find the median of the values, and then replace each value by A if it is above the median and B if it is below the median. Then apply the runs test to the resulting sequence of A's and B's. What does the result suggest about the stock market as an investment consideration?

943 985 969 842 951 1036 1052 892 882 1015 1000 908 898 1000 1024
1071 1287 1287 1553 1956 2722 2184 2791 3000 3169 3413 3794 3978 5667 6624

With the data in L1, highlight L2 and let L2=L1−median(L1, as in the bottom line of screen (32) with median from ↑LIST<MATH>4. Press **ENTER** for screen (33).

It is now easy to detect the runs of negatives (below the median) and positive (above the median) in L2.

Use the ▼ key with L2 to find that the first 15 values are negative and the last 15 values positive for only two runs (see screen (33)). With G = 2, n1 = 15, and n2 = 15, Table A-10 gives the cutoff values of 10 and 22 for $\alpha = 0.05$.

We reject the null hypothesis of randomness above and below the median. "The lack of randomness is due to an upward trend, which makes the stock market an attractive investment consideration."

Appendix

Loading Data and Programs from a Computer to a TI-83, and from One TI-83 to Another

A disk to accompany this companion is available from the publisher in both PC and Macintosh formats and contains the following files:

1. The data sets from Appendix B of *Elementary Statistics* (7th ed.) by Mario F. Triola are given in ASCII format; see the ASCII subdirectory or folder. The file names of the data are the same as those listed as STATDISK variable names in Appendix B of the text (e.g., Data Set #1: HHSIZE, METAL, PAPER, PLAS, . . ., TOTAL).

2. Two programs, one named A1ANOVA.83p (used in Chapter 11) and one named A2MULREG.83p (used in Chapter 9).

Your instructor will probably load the programs into your TI-83 if and when you use them and perhaps transfer some of the larger data sets into your TI-83 when you are ready to work with them. In this case, you may still want to investigate part B on page 116, on sharing data from your TI-83 with a classmate's TI-83 or TI-82.

A. Loading Programs or Data from a Computer to a TI-83

Warning: Open a program file and click on the protect (or lock) option before loading that program to a TI-83. This eliminates the possibility of inadvertently altering the program in the TI-83 requiring you to reload the original program.

1. Connect your TI-83 to your computer with the TI-GRAPH LINK cable. The TI-83 link port is located at the center of the bottom edge of the calculator.

2. Turn on your TI-83 and leave on the Home screen.

3. (a) **For A1ANOVA.83p or A2MULREG.83p:** On the computer with the TI-GRAPH LINK software, open the SEND menu and select the programs from their storage location. Move along the file names for A1ANOVA.83p and/or A2MULREG.83p, add the program(s) to be transmitted, and then SEND to your TI-83.

   (b) **To Load Data from an ASCII file**: Before opening the Send menu as shown previously, you must open the Tools menu and select "Import ASCII DATA". Then import to the desired ASCII file, which adds a ".83L" after the name. Open the SEND menu, select the list(s) to be added, and send to your TI-83. You can view the list on your TI-83 with ↑LIST<NAMES>. The Macintosh Graph Link saves a list as LIMP. Paste this to the Home screen and store with a name of your choice (e.g., LIMP STO►AGE) so that the next list transferred can be stored over LIMP.

See the TI-GRAPH LINK GUIDEBOOK for full details, or refer to page B-12 and B-13 of the guidebook that came with your TI-83 for information on where you can find assistance with the TI-Graph Link.

B. Loading Programs or Data from One TI-83 to Another TI-83

1. Connect one TI-83 to another with the cable that came with the calculator. The TI-83 link port is located at the center of the bottom edge of the calculator.

2. Press ↑**LINK** on the TI-83 that is to receive the program or data, and then press ▶ to highlight **RECEIVE**. Press **1** or **ENTER** to select Receive and have the message "Waiting..." displayed.

3. Press ↑**LINK** on the TI-83 that is to send the program or data, and then press **3** for **Program** or **4** for **List**. Use ▼ to highlight the program or list name to be sent, and press **ENTER**. A small darkened box will appear in front of the name (easier to see if you move the cursor pointer down). Repeat the process for all programs or lists to be sent. Press ▶ to highlight **TRANSMIT**. Press **1** or **ENTER** to Transmit the programs or lists.

Note: The only data type you can transmit from a TI-83 to a TI-82 is list data stored in L1 through L6.

See your TI-83 GUIDEBOOK (Chapter 19) for complete instructions.

Index

TI-83 Quick Reference

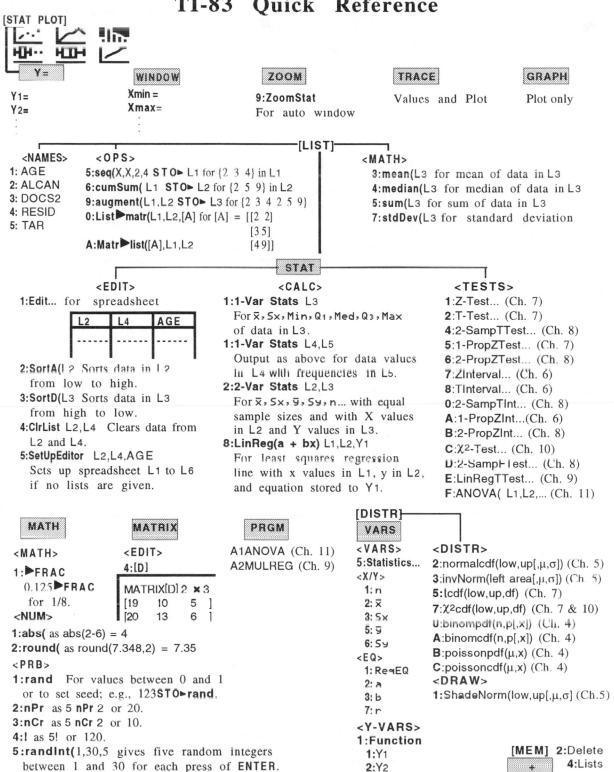

[STAT PLOT]

Y =
Y1=
Y2=

WINDOW
Xmin =
Xmax=

ZOOM
9:ZoomStat
For auto window

TRACE
Values and Plot

GRAPH
Plot only

[LIST]

<NAMES>
1: AGE
2: ALCAN
3: DOCS2
4: RESID
5: TAR

<OPS>
5:seq(X,X,2,4 **STO►** L1 for {2 3 4} in L1
6:cumSum(L1 **STO►** L2 for {2 5 9} in L2
9:augment(L1,L2 **STO►** L3 for {2 3 4 2 5 9}
0:List**►**matr(L1,L2,[A] for [A] = [[2 2]
                                    [3 5]
A:Matr**►**list([A],L1,L2   [4 9]]

<MATH>
3:mean(L3 for mean of data in L3
4:median(L3 for median of data in L3
5:sum(L3 for sum of data in L3
7:stdDev(L3 for standard deviation

STAT

<EDIT>
1:Edit... for spreadsheet

L2	L4	AGE
------	------	------

2:SortA(L2 Sorts data in L2 from low to high.
3:SortD(L3 Sorts data in L3 from high to low.
4:ClrList L2,L4 Clears data from L2 and L4.
5:SetUpEditor L2,L4,AGE Sets up spreadsheet L1 to L6 if no lists are given.

<CALC>
1:1-Var Stats L3
For $\bar{x}, Sx, Min, Q_1, Med, Q_3, Max$ of data in L3.
1:1-Var Stats L4,L5
Output as above for data values in L4 with frequencies in L5.
2:2-Var Stats L2,L3
For $\bar{x}, Sx, \bar{y}, Sy, n$... with equal sample sizes and with X values in L2 and Y values in L3.
8:LinReg(a + bx) L1,L2,Y1
For least squares regression line with x values in L1, y in L2, and equation stored to Y1.

<TESTS>
1:Z-Test... (Ch. 7)
2:T-Test... (Ch. 7)
4:2-SampTTest... (Ch. 8)
5:1-PropZTest... (Ch. 7)
6:2-PropZTest... (Ch. 8)
7:ZInterval... (Ch. 6)
8:TInterval... (Ch. 6)
0:2-SampTInt... (Ch. 8)
A:1-PropZInt...(Ch. 6)
B:2-PropZInt... (Ch. 8)
C:χ^2-Test... (Ch. 10)
D:2-SampFTest... (Ch. 8)
E:LinRegTTest... (Ch. 9)
F:ANOVA(L1,L2,... (Ch. 11)

MATH

<MATH>
1:**►FRAC**
0.125**►FRAC**
for 1/8.
<NUM>
1:abs(as abs(2-6) = 4
2:round(as round(7.348,2) = 7.35
<PRB>
1:rand For values between 0 and 1 or to set seed; e.g., 123**STO►rand**.
2:nPr as 5 nPr 2 or 20.
3:nCr as 5 nCr 2 or 10.
4:! as 5! or 120.
5:randInt(1,30,5 gives five random integers between 1 and 30 for each press of ENTER.

MATRIX

<EDIT>
4:[D]

MATRIX[D] 2 **×** 3
[19 10 5]
[20 13 6]

PRGM
A1ANOVA (Ch. 11)
A2MULREG (Ch. 9)

[DISTR]
VARS

<VARS>
5:Statistics...
<X/Y>
1: n
2: $\bar{x}$
3: Sx
5: $\bar{y}$
6: Sy
<EQ>
1: RegEQ
2: a
3: b
7: r
<Y-VARS>
1:Function
1:Y1
2:Y2

<DISTR>
2:normalcdf(low,up[,μ,σ]) (Ch. 5)
3:invNorm(left area[,μ,σ]) (Ch. 5)
5:tcdf(low,up,df) (Ch. 7)
7:χ^2cdf(low,up,df) (Ch. 7 & 10)
0:binompdf(n,p[,x]) (Ch. 4)
A:binomcdf(n,p[,x]) (Ch. 4)
B:poissonpdf(μ,x) (Ch. 4)
C:poissoncdf(μ,x) (Ch. 4)
<DRAW>
1:ShadeNorm(low,up[,μ,σ] (Ch.5)

[MEM] 2:Delete
       4:Lists